Simba Chai

The Kenya tea industry

MICHAEL MCWILLIAM

Editing, typesetting and design by
Prepare to Publish Ltd
www.preparetopublish.com

CONTENTS

PREFACE

The opportunity to research and write about Kenya's tea industry came about as a result of being elected to a research studentship at Nuffield College, Oxford in 1955.

My motivation for extending my time in Oxford for another two years was not entirely academic, but there was a serious element to my research, allied to the fact that I had spent my childhood on a tea estate in Kericho. It was this that led me to the idea of undertaking a study of the tea industry for a postgraduate thesis degree that the university offered in those days, known as a B.Litt. The Tea Board of Kenya agreed to facilitate my research, but it exacted a condition, accepted by the university, that it could control public access to the finished product. I was awarded the degree in 1957 but, unexpectedly, the Tea Board decided to enforce its veto. Thereafter, the Bodleian Library in Oxford denied reader access to the thesis and, naturally, there was no question of its publication. Some 50 years later, in retirement now, I took up the old document with a view to salvaging something and updating the narrative, particularly with regard to the smallholder tea sector, about which I had been somewhat dismissive in my thesis. I had recently served on the board of the Commonwealth Development Corporation (CDC), and written its history, and had become very much aware of its important role in promoting smallholder agriculture, and of its flagship project, the Kenya Tea Development Authority (KTDA). Two research visits to Kenya in 2003 and 2004 provided a mass of information on the emergence of tea as the country's principal export earner. However, there was another lengthy pause for domestic reasons before the project was completed.

Some of this book is based on archival research through which I was able to see material that may no longer be readily available, if at all. The narrative of the early development of tea growing has benefitted from access to private memoranda and recorded memoirs, while the appraisal of developments in tea cultivation and manufacture is the product of the research visits referred to. The Bibliography provides more detail on the sources. The turn of the 20th century was an appropriate point to conclude the historical narrative, but I have taken the opportunity to convey a sense of where the industry is today, to point up some of the contemporary issues confronting it and to hint at the direction it might take in the future.

My father was recruited by Brooke Bond to go out to Mombasa in 1926 as a trainee salesman. The following year he was sent to Kericho to take part in the opening up of Brooke Bond's tea estates there and worked on Cheboswa estate for several years until later returning to tea marketing. He married his childhood sweetheart on his first leave in 1930 and, as recounted in Chapter 5, spent the 1930s putting in place the tea cartel to manage tea supplies to the domestic market. During World War Two he managed the bulk tea contracts to the Ministry of Food on secondment to the government as the East Africa Food Controller. He was responsible for tea marketing in East Africa until his retirement in 1958, but stayed on in Kenya for another ten years, serving as Secretary of the Tea Board, before returning to Gloucestershire. I dedicate this book in his memory.

I also record my gratitude to the many people in Kenya and Britain who shared their knowledge and memories and who worked, or had worked for, tea organisations and companies. These include James Finlay/Swire, Brooke Bond/Unilever, Eastern

Produce/Linton Park, the KTDA, Kenya Tea Board, Kenya Tea Growers Association, Kenya Tea Packers, East Africa Tea Brokers, Combrok, Thompson Lloyd & Ewart and the International Tea Committee. Some of them warrant individual mentions for their helpfulness and insights. In Kenya: Nick Paterson, on retirement from James Finlay, served as visiting agent to both his old company and Brooke Bond; Nev Davies, chief executive of James Finlay in Kenya during my visit; Richard Fairburn, chief executive of Unilever at the time; Titus Korir, deputy to Nev Davies; Stephen Musongo, estate manager of Kaproret estate; Eric Kimani, chief executive of the KTDA; Francis Wacima, tea farmer at Limuru; John Nottingham, former district officer and tea farmer; and Joyce Pickford, widow of a Kericho manager. In Britain: Oliver Brooke, former chairman of Brooke Bond Liebig in Kenya; Tom Brazier, former director of Brooke Bond Liebig; Chris Tyler, former Brooke Bond factory manager in Kericho; Lindsay Stone-Wigg, former superintendent of James Finlay; Rupert Hogg, managing director of James Finlay; and Duncan Gilmour, secretary of James Finlay.

And finally, this book would not have achieved its present form without the constructive and ever-helpful advice of my editors James Attlee, Derek Collett and Andrew Chapman at Prepare to Publish, and the guidance of Professor David Anderson with regard to referencing and preserving the archive material.

INTRODUCTION

When the consumption of tea spread beyond the caravan and shipping routes of East Asia and started to be shipped in quantity to Europe in the 18th century, it could be claimed that a global product was being established. Tea became embedded in the consumer cultures of Europe and Russia and throughout the Islamic world. India not only became the leading producer of tea for export, but the product became widely adopted in the nation itself as a staple drink. The 20th century saw its further penetration into the Americas and Africa, and the emergence of Kenya as a key player in the industry. Societies have taken to tea drinking in contrasting ways, ranging from the formal rituals of the Japanese tea ceremony to the casual office or factory tea break in Great Britain or the thriving tea stalls in South Asian street markets. In these markedly different fashions, much of the world has benefitted from tea's unique combination of taste, a mild caffeine stimulant and the safety factor of being a beverage made using boiled water. To this has been added a more general acceptance that its properties are beneficial to health.

Kenya is justly famous for its wildlife and for the tourism sector of its economy. Less well known is that it is now the largest supplier of tea to world markets, ahead of India, China and Sri Lanka, and tea is by far the country's most valuable commodity export, accounting for around one-fifth of total exports by the beginning of the 21st century. This is reflected in the proportion of the country that has been dedicated to tea production. In 1945 this totalled 6,400 hectares (ha). Over the following 60 years it expanded rapidly: by 1965 it had quadrupled and by the year 2000 it had increased by a factor of 25 to reach 161,000 ha. This remarkable growth was the result of two distinctive aspects of colonial rule. On the one hand, Kenya was a British colony in which enterprising settler farmers were experimenting with the cultivation of new crops and tea plantation investors were extending their territory beyond India and Ceylon; on the other, the colonial administration in East Africa concentrated on bringing about managed development, perhaps the most successful example of which was the creation of the smallholder tea sector. As a consequence, the tea industry today comprises two contrasting modes of operation: a plantation sector largely owned and managed by large international companies with 35,000 ha of tea plantations, which constitute the biggest employers in agriculture; and a smallholder sector of more than 450,000 Kenyan farmers with 126,000 ha of tea, who are themselves a major source of employment. In between these two poles there are a number of privately owned tea estates, many of them properties that were owned by earlier settlers but then acquired by elements of Kenya's new political and economic elite.

When the East Africa Protectorate was declared in July 1895, there was no clear view of the potential of the territory that would become known as Kenya to generate wealth for its European settlers, nor was there any obvious commodity suited to export, apart from ivory. The country's cultivators and pastoralists lived in societies that were largely self-sufficient. The climate of the highlands behind the coastal plain suggested that there were prospects for farming and settlement and this became the economic foundation of future development. With much trial and error over a 50-year period, a diversified agriculture evolved in Kenya. Export revenues were earned from coffee, sisal, pyrethrum and tea, and the country became self-sufficient in food grains, meat and dairy produce to feed a growing and increasingly urbanising population. Since independence, two important earning streams have been added to the economy: air-freighted flowers and vegetables; and a tourism sector based on the country's wildlife reserves. Kenya became a regional manufacturing and services centre for East Africa before independence, and this role has been much enhanced by the choice of Nairobi for regional headquarters by numerous international agencies and many other organisations. Nevertheless, it is still the case that agriculture is fundamental to Kenya's economy. As the nation entered the 21st century, around half its GDP was accounted for by its agriculture and around half the country's commodity exports comprised food, beverages and tobacco.

The misperceptions over traditional land tenure in tribal societies at the time the colony was established were reinforced by the results of recent conflict and these led to a view that much of the highlands were vacant or so sparsely populated that incoming land occupation and settlement could take place. The drive to acquire high-quality agricultural land was driven by the need to defray the costs of the new administration. Conflicting views on the matter were a dominant feature of politics almost up until the attainment of independence in 1963. The tea plantation developments of the 1920s and after were only made possible by the land grants that took place in the early 20th century, alongside the allocation of farmland for settlement. This process was completed by the 1950s, although some adjustments of ownership have continued since then. An attempt was made in the 1930s to rectify the misjudgements through the Carter Land Commission, and again in the early post-war period. Another factor that was critical to early progress in the colony, and also the source of much controversy, was procuring a workforce from tribal communities that were engaged in subsistence agriculture and barely involved in the cash economy. Tea cultivation is labour-intensive, so attracting a workforce and retaining it were major preoccupations of the tea companies as they wrestled with the goal of establishing a settled labour force on their estates. Towards the end of the colonial period, and as part of a wider, ambitious initiative to transform the economic prospects of

Kenyan farmers, plans were drawn up and implemented by the Department of Agriculture to make it possible for small farmers to grow tea to a high standard and to have it processed in order to provide their main source of income. A state corporation was established, the Kenya Tea Development Authority, to manage the project, which attracted the support of the Commonwealth Development Corporation and the World Bank. It was a great success, to the extent that smallholder tea production has overtaken that of the plantation sector and has propelled Kenya to its leading position in the global tea trade.

The way in which the tea industry evolved in the 1930s was shaped by the politics and rules of an international commodity scheme that operated between 1933 and 1948. The International Tea Restriction Scheme was a response to falling tea prices and was an attempt to rectify this by controlling the quantity of tea exported to world markets, and by stopping new planting. Despite continuing concerns about imbalance between world supply and demand for tea, and a predisposition of the Food and Agriculture Organization (FAO) for many years to support new measures to control the market, not to mention the alluring example of the Organization of the Petroleum Exporting Countries (OPEC), it has not proved possible to re-establish a new tea restriction scheme and thus Kenya was able to continue to expand.

Tea as a beverage was unknown in East Africa before the colonial period, except amongst Arabs at the coast. But the potential for a new market was spotted by Brooke Bond's Indian marketing organisation and sales were made to Mombasa in the early years of the 20th century. The International Tea Restriction Scheme reinforced the importance of stimulating tea consumption in countries where tea was grown; in Kenya, this gave rise to an initiative to establish a producers' cartel to supply the domestic market. The East Africa Tea Agreement, or Pool as it came to be known, was set up in 1938 under the operational management of Brooke Bond. It continued for 40 years and successfully established East Africa, and especially Kenya, as a tea-drinking society, and this is the only part of tropical Africa where this is the case.

After the end of World War Two there was a large expansion of the estates sector from 6,500 to 35,000 ha by the end of the century, and this has since grown to over 90,000 ha. This investment was undertaken both by the established tea companies that had been held back by the International Tea Restriction Scheme, and also by a wave of investment from tea companies with long-standing operations in India and Sri Lanka who were hedging political risk. Meantime, the newly established smallholder sector encompassed 91,000 ha by the end of the century and has since expanded further to 141,000 ha. All but 10 percent of the output from these vast developments was exported to world markets. Kenya has been very fortunate that this increased supply to world markets has not been disruptive. This is mainly on account of the fact that India and China have retained a very high proportion of their own tea production to satisfy rising domestic consumption in their own countries.

Tea had been grown and manufactured in India for 80 years before development got under way in Kenya, so that cultivation and processing methods were well tested and established and were readily transposed to the new estates. After the war, advances in understanding of plant physiology, fertilisers and herbicides began to be applied to the tea industry, along with engineering developments and insights into the chemical changes taking place during manufacture. These developments transformed crop yields, labour productivity, the development of new consumer products and quality control.

Kenya has remained welcoming to foreign investment under successive governments, unlike a number of African countries in the decades following political independence. However, international opinion has been critical of the conduct of domestic politics and of the prevalence of serious corruption in public life, which has damaged more vulnerable parts of the economy and the public services. The tea industry has been relatively unharmed by these events, except for the tragic ethnic pogroms that took place in western Kenya after the 2008 elections; otherwise, it could be said that the industry entered the 21st century in good shape. The political economy of tea has been helpful in this regard. The smallholder sector is large and vocal and its farmers span several ethnic groups. Kenyans who have acquired tea properties are well connected to the country's ruling elite. The major tea companies set high standards in employment and social responsibility and are crucially linked to the export markets for tea. The rapid end of colonial rule in the early 1960s caught the companies unprepared as regards their management cadres, and their responses varied. But the estates were all substantially managed by Kenyans well before the turn of the century, with a high level of proficiency. Taken together, these three elements provide the industry with a measure of security and stability.

The tea industry faced a number of challenges at the commencement of the new century, arising in part from its very success. The wide gap in output between the most productive units, which yield around 6,000–8,000 kg of tea per hectare, and the average yield of less than half that amount represents a huge loss in income potential, especially for the smallholder sector. For the estates, the priority is to improve productivity and maintain profitability in the face of rising labour costs and weak global market prices. For smallholders, the aspiration of achieving increased yields and profits is, above all, a challenge for the advisory role played by the Kenya Tea Development Agency (as it is now called) and its very large constituency of farmers. The tea factories, which are now owned by the smallholders, face the issue of how best to benefit from – and finance – advances in manufacturing practice.

The goal of seeing stabilised communities working on the estates has been undermined by ethnic tensions and by the scourge of HIV. But the Kenyan industry is fortunate in having the presence of international investors with a strong commitment to their social responsibilities, and an indication of their approach to resolving these problems is given.

As a result of Kenya's huge production of tea for export, domestic consumption only accounts for around 10 percent of its output. This contrasts with India, which absorbs around 80 percent of its own production. The Kenya Tea Packers Association has a challenge to promote tea drinking more widely in its domestic hinterland in the way pioneered by Brooke Bond in the early post-war era, since this is unlikely to be a priority for the major international companies. Tea drinking has been strongly embedded in the culture of many diverse societies and Kenya's example indicates that it could spread more widely in Africa. In contemporary Western urban culture, the drinking of tea is less strongly promoted than that of coffee, or of manufactured drinks like Coca-Cola and Pepsi. The global tea industry could do with the innovative flair of a Starbucks to re-establish its place as a beverage of choice with a younger generation and the widening appreciation of its health benefits indicates a possible way forward. Meantime, Kenya has consolidated its place as an essential supplier to the tea drinkers of Europe, North America and the Middle East.

1

HOW IT ALL BEGAN

From China to Africa

Tea was introduced in England in the mid-17th century as an exotic beverage for the well-to-do and its popularity spread very quickly. For a rapidly urbanising population, for whom access to safe drinking water was a public health problem, boiled water flavoured with tea, the addition of milk and a liberal helping of sugar (now readily available from the West Indies) replaced beer as the everyday beverage and was nutritious to boot. Tea became a national drink from the drawing room to the factory canteen, and the source of supply was a single port in southern China. Canton was up to a year's voyage away and the trade was strictly regulated by the Chinese authorities, so that for a long time very little was known about how it was grown or manufactured. The explosive demand for the product led to much adulteration, both in China itself (the most famous lineal descendant being the bergamot-adulterated Earl Grey tea) and in the English trade, in order to offset its high price. A lively smuggling trade developed in order to avoid customs duties, which provided more opportunities for innovative blending. Famously, the attempt by Lord North's government to assert its right to tax the American colonies by imposing a tea duty was the spark that led to the War of Independence.

China's supply of tea for Europe was based on peasant production and a primitive porterage supply line that could not respond quickly to rising demand. The East India Company controlled all trade with China, where ordinary merchandise trade was highly restricted, so that payment for tea exports had to be made in silver. The solution found by the Company during the 18th century was to promote a vast smuggling trade in opium from Bengal. The opium was shipped clandestinely to China and sold for silver at great profit, which in turn provided the wherewithal to pay for the Company's tea purchases. The varied legacy of this fateful trade included the Opium Wars, the acquisition of Hong Kong, the opening of the treaty ports and their associated 'unequal treaties', which have continued to reverberate through to the present day.

The East India Company's monopoly of the China trade was terminated in 1834, and that same year a Tea Committee was set up in Calcutta to examine the possibility of establishing a tea industry in India, where the beverage was unknown. A mission was despatched to China to collect tea seed. Curiously, specimens of the camellia tree had been sent to the Calcutta Botanical Garden over many years, but there had been a reluctance to conclude that this was the same species as in China. In the event, the Tea Committee did pronounce that tea was indigenous in Assam and, over the ensuing years, plantings were made of both Assam and China tea. The first samples were sent to London for auction in 1839 and the excitement this generated led to the founding of the Assam Company, the first of many Indian tea companies. In 1848 the East India Company recruited the notable plant hunter Robert Fortune, manager of the Chelsea Physic Garden, to return to China to make a further collection of tea plants and seeds. It was on this trip that Fortune solved the 100-year puzzle of green and black tea, by establishing that the difference lay in its manufacture and was not the product of different plant varieties. Rapid expansion of the industry led, by 1888, to tea imports to England from India exceeding those from China for the first time.[1]

Meanwhile, the first tea planting had taken place in Ceylon in 1867, where it was found to be a successful replacement for coffee, which had been devastated by a fungal disease. By the end of the century Ceylon was a major tea exporter, vying with India, and with numerous corporate links between the plantation companies. Just as in Assam, the tea plantations were established in forested areas of the country and large numbers of workers had to be induced to come to live there: from elsewhere in Bengal in the case of Assam, and from Tamil Nadu in south India in the case of Ceylon.

The next step in the westward migration of *Camellia sinensis* to Africa followed closely on the establishment of the East Africa Protectorate in 1895 and the arrival of the early settlers. In their search for potential agricultural crops, some of them had personal connections with India and, not long after, visitors from Ceylon and India spotted the potential for tea.

Early Settlers and the Crown Lands Ordinance

The establishment of British rule in Kenya presented an immediate economic challenge: the administration was burdened with the task of servicing the loan costs of the railway from Mombasa to Kisumu on Lake Victoria, 580 miles into the interior, and making it viable; there were no obvious tropical export commodities or signs of surplus production; the small population of the country had recently been ravaged by disease and warfare; and British rule itself was challenged for several years in the west of the country, requiring numerous punitive and expensive military engagements. On the other hand, early explorers and visitors noted the suitability of the highlands for European settlement, and this fitted with the ethos of 19th-century imperialism where it had been such a feature, as highland settlement had been such a feature of the development of Canada, Australia, New Zealand and also South Africa. No surprise therefore that Sir Charles Eliot, the commissioner, encouraged immigrant settlers to the highlands with a view to stimulating development and government revenues. Although the initial emphasis was on the Rift Valley for ranching and cereal crops, land grants were also made in districts that subsequently became significant for tea growing: Limuru, Kericho and Nandi.

In Limuru, the government began allocating land in 1903, under pressure from the flow of new settlers. Its authority was the 1901 East Africa (Lands) Order in Council, which defined Crown land as all public land controlled by His Majesty, and the Crown Lands Ordinance of 1902, which restricted the commissioner from selling or leasing land that was in actual occupation by 'natives', although he could do so if occupation had lapsed. This proved to be a fallible basis of operation for three reasons. First, there had been a severe reduction in the Kikuyu population in the recent past as a consequence of a devastating smallpox epidemic, so that areas that had been in regular occupation were currently vacant. Second, there was little understanding among the colonial administrators at the time of the complex system of Kikuyu land use and ownership.[2] The Kikuyu lived in dispersed family households rather than in villages, on land owned by family groups and inheritable within them. They cultivated scattered plots over a ridge and valley, identifiable to those concerned, and practised a system of shifting cultivation. Much of the area had previously been under the control of a different tribe, the Dorobo, and there was a history of both purchase and conquest between different tribes. Tenancy arrangements with non-family persons were not uncommon. All this produced a situation in which land that appeared vacant might actually be subject to a history of cultivation and ownership.

Thirdly, onto this scene now came a flow of settlers, many with experience of farming in South Africa, impatient to acquire farms and frustrated by dealing with an understaffed Lands Department. No official survey had been made to define precise areas that were to be made available for settlement in terms of the Ordinance. In practice, settlers selected their farms and then registered them, notwithstanding the presence of Kikuyu farmers on parts of the land, subject only to a requirement to pay compensation – on a uniform tariff for occupancy rights. There were numerous instances of non-compliance in this regard.

Eliot was in favour of the interpenetration of settlers with the Kikuyu cultivators, rather than the demarcation of reserves, and the outcome was much alienation to settlers of land on which the original farmers were still present. Under the terms of the Ordinance they had no legal status, other than compensation, and those remaining were deemed to be squatters; however, this suited the settlers since it offered a potential source of farm labour, and the practice of resident families cultivating their own smallholdings but providing labour for the farm owner became widespread in the settlement areas in the highlands. Although the Crown Lands Ordinance gave the government control over land allocation to settlers and for public purposes, its disregard for traditional systems of land usage, and its eagerness to create opportunities for commercial agriculture, stored up serious political problems for the future. Some of these problems were addressed in the 1930s in the report of the Kenya Land Commission, but others still fester.

In the early 1900s, Kenya represented a new frontier for a number of farming families in South Africa and there was a wave of settlement, especially to the high plateau around Eldoret. Others obtained land grants in areas that eventually were shown to have tea potential: in Limuru (Caine), Kericho (Wilson) and Nandi (Turton). There are two main sources of information on the early tea-growing settlers. Hugh Thomas was one of the first planters to be transferred from India to Kenya in 1927; he collected reminiscences in the 1930s and subsequently prepared an account of the early days for the benefit of the newly established Tea Board of Kenya in 1950, and he also subsequently wrote a history of his company. John Wilson's father was allocated a farm near Lumbwa in 1911 where John grew up; he set down his recollections of the first farm allocations in the Kericho area.[3] The story of the Caine brothers, one of whom had been in India, is the best known. They were allotted a 3,000-acre freehold property, which they called Caineville estate, and where they inherited 374 squatters and 30 permanent workers. In 1903, GW Caine obtained tea seed from Madras. By 1908 he was producing hand-made, sun-dried tea. A sample was sent to the Imperial Institute in London by the Department of Agriculture, and a tea broker's report was obtained.[4] The conclusion drawn by the Director of the Imperial Institute was prophetic: "The investigation showed that tea of good saleable value can be grown in the Limuru district of the East Africa Protectorate, with a good prospect of success."[5] A number of the early bushes – now mature seed-bearing trees – are still on the site.

The initiative for planting the first tea bushes in Kericho was taken by the first district commissioner, Partington, while he was stationed there between 1903 and 1905. He was so enthusiastic about the prospects that he was said to have applied for a grant of land on retirement in order to grow tea; however, land alienation had not yet commenced in the district and he was refused.[6] The mature trees remained in the district headquarters, but were unfortunately cut down in the early 1950s in order to make way for a new police station. The alienation of farms commenced near Lumbwa in 1907, to Clift at Kapsongoi and Wilson at Chisimot, and proceeded through Kericho and towards Sotik. Wilson recalled 15 settlers in his memoir of the early days.[7]

In 1912, Barclay at Kapkorech imported tea seed from Ceylon. The farm was acquired by Mathews in 1919 and he began to extend the area under tea, so that by 1924 there were 50 acres planted. In that year he built a small tea factory, which became the first in Kenya. The Orchardson brothers acquired a farm at Kimugu in 1910 and Ian Orchardson later put down his recollections of the somewhat haphazard early initiatives:

"In 1912 or 1913 we imported a maund of seed through a friend – poor jat, poor seed – having read in Economic Products of India how to plant tea we left it for 18 months and planted it out. It all died except a few poor things we left in the nursery. A few are still there. Some I gave to Grice and Howland which grew there for some time... I sent samples of sun-dried tea to England in 1916 and got a rather encouraging report but could interest no capital. We always made and drank our own tea from 1916 onwards."[8]

In 1919, Orchardson tried to form a syndicate to acquire a larger land grant at Kitumbe, but this was unsuccessful as the government had other plans for the area. A syndicate headed by Butterfield (including Birkitt and Brierley), which owned land at Jamji that had originally been planted to flax and coffee, established an experimental tea plot in 1921. This property was purchased by Lord

Egerton, who went on to develop a full tea estate, as well as acquiring the Kapkorech estate from Mathews. The Butterfield syndicate started planting tea on another property at Chemosit, which by 1924 had reached 150 acres, and further tea planting was commenced at Buret. The two properties were combined to form the Buret Tea Company, which had 1,200 acres of potential tea land; 375 acres had been planted by 1927 and 920 two years later, and the land was fully planted by 1934. In 1926 the property that became the Mau Forest estate was purchased by Grant and developed as a tea estate with the support of one of the incoming tea companies from India, James Finlay. Also in 1926, a syndicate led by Colonel William Brayne, the Gorwat Syndicate, purchased the nearby Kaisugu property and developed a 500-acre estate.

The commencement of tea development in Kericho in the 1920s naturally stimulated interest amongst farmers who had settled in the buffer zone of Sotik, between the Kipsigis and Kisii peoples, and a first planting was made by Mathews. The introduction of the Tea Restriction Scheme in 1933 prevented any further development until the Department of Agriculture took up the cause of failed coffee farmers at the time of the scheme's renewal in 1938. The South Sotik Settlers Association noted that there were four experimental tea plantings when applying for 1,000 acres, and the North Sotik Farmers, applying for a similar amount, stated "The applicants are in a position to finance the industry and, in the event of a grant being made, they will form themselves into a company or cooperative society which would build a central factory for the manufacture of the tea."[9] In the event, only 330 acres were awarded to 11 farmers in the combined district: Jan George developed Arroket to a small factory estate, taking in leaf from other growers, including Maitland Edye on Monieri, Preston on Kaptembwe, Shaw at Kipkebe and Dawson at Kimoro. A precedent had been set for more ambitious development of the tea potential of Sotik, which took place post-war.

The concept behind the Crown Lands Ordinance, which gave recognition to African land rights only where cultivation was taking place with effective occupation, was hard on the pastoral Kipsigis people in the Kericho area on two counts: firstly, the Ordinance did not recognise occasional cultivation and pastoral use of land and, secondly, it did not allow for the recent withdrawal of Kipsigis from areas of historic occupation as a result of warfare. The Kipsigis people came to Kenya from the Sudan in the 18th century and occupied the Londiani area initially, where the hill Tiluap Sigis is still a place of special significance.[10] Their arrival led to conflict with the Masai and with the Kisii people further west. 1890 was the occasion of the last major battle at Saosa in Kericho, in which the Kipsigis were defeated with heavy casualties. The consequence was that the district was sparsely populated when the first colonial administrative post was established in 1902. The government's concern at that time was to dissuade the Kipsigis from giving support to their 'cousins' the Nandi, who had been in conflict with it over the caravan route – and later railway line – to Uganda. A punitive expedition against the Kipsigis was deemed necessary in 1905 for having raided the Masai, with cattle confiscation and hut destruction, and a buffer zone of settler farms was created in Sotik.[11] The farms allocated in Sotik had the political purpose of forming a barrier between the Kisii and Kipsigis people, but they happened to be in a traditional Kipsigis grazing area. Most of the farms were not occupied initially, so that the Kipsigis were able to continue running their stock, but problems arose once the farms began to be developed.

The Kipsigis position, essentially, as put on record before the Carter Land Commission in the 1930s, was that all Kericho was Kipsigis country since they had moved into the area in the previous century. Kericho itself was the name of a prominent clan and *laibon*. Early administrative officers recorded their views unambiguously; thus the provincial commissioner, RW Hemsted, wrote in 1925: "I have always held the opinion that the Lumbwa have not been fairly treated as regards land, with the result that many thousands of them have been forced to go onto the farms to find grazing for their cattle. Many of the farms should never have been alienated."[12] The Carter Commission published its final report in 1934, having collected voluminous evidence on Kenya's land issues over the preceding two years.[13]

The district commissioner for Kericho, Tompkinson, gave written evidence: "If one were to consider for a moment that [the Kipsigis] had rights to any land once occupied by them, then their claims would be out of all proportion to their needs, nor would there be any justification to support such claims on economic grounds. However, actual alienation was harmful."[14] He enumerated the Chematum salt lick that had been alienated in order to make it easier for the settler farmers to attract resident labour ('squatters') and, in Sotik, "the Kipsigis were deprived of practically all the best of their grazing land" in order to institute a buffer between them and the Kisii, with whom there was a history of tribal warfare. However, much of this land had not actually been developed.

The Carter Commission concluded (paragraph 1149): "The Lumbwa claim that a great proportion of their grazing grounds have been taken away from them and alienated to Europeans. This is undisputed." In recompense, the Commission proposed first, that the people should regain use of Chematum salt lick and also access to the Kibert Litein salt lick and second, that an area of 148 square miles at Chapalunga be added to the native reserve. It made no mention of the Kericho problem. Interestingly, one of the Kericho farmers, Ian Orchardson, who had lived there since 1911 and was an authority on the Kipsigis, himself submitted a memorandum to the Commission on the alienation of Kericho land: "This land has been occupied by the Kipsigis, who were alternatively turned off and taken on as squatters by subsequent owners... I think we should remember that much of this 'forest' was occupied by the Kipsigis until the arrival of European Government. In fact this was the area first occupied by them in this district 180 years ago."[15] Orchardson offered to give oral evidence, but he was not called. His solution in Kericho was that an area of Crown land, known as Kimulot and amounting to 12,000 acres, should be returned as compensation. It was already subject to partial, unauthorised, occupation and had been forcibly cleared in 1922. His proposal was not taken up, and the government again cleared the area in 1936. Kimulot came to prominence in the post-war period.

At the end of the First World War the Kenya government strongly supported a scheme to recruit former army officers as settler farmers, the notion being that, with proven management abilities for the purpose of supervising an African labour force and a modicum of capital, a class of gentlemen farmers could be established in the colony. Plans were approved for 1,000 settlers and a million-acre scheme. A number of these settlers went to Nandi district. Against this background, a promoter of dubious credentials won the support of the Colonial Office in 1919 for another project to settle 55 disabled officers in a farming cooperative in Kericho to grow flax. The project had the support of the Empire Flax Growing Committee, which was seeking new sources of flax after the loss of Russian supplies, and it was known that a number of Kenya settlers had successfully started growing flax. Each settler was required to put up £500; the British government provided an initial training grant and a temporary increase in pension income. The government of Kenya made available to the British Disabled Officer Colony 25,000 acres that had been excised from the reserve in 1909 and designated as Crown Land.

The first 13 officers arrived at the end of 1919 and the remainder in the following year. The local provincial administration built an access track and provided some advisory help. Unhappily, the price of flax collapsed in 1920 from £580 a ton to £80 a ton. By the end of 1921, 800 acres of flax had been planted and 200 acres of maize. But it was a year of drought and the scheme was in debt to its members and the bank. The Kenya government converted the land titles to freehold for nil consideration to improve security, and the Colonial Office was supportive of attempts to raise fresh capital in England (declined by the Treasury), and the Officers Association lent £10,000. However, in late 1922, the scheme had to be put into liquidation with debts of £62,000, and the government was urgently seeking buyers for the land.[16]

This, then, was the context in which Brooke Bond and James Finlay, two companies with tea plantation interests in India, became aware of the potential for tea growing in Kenya and then committed themselves to realising it.

Arrival of Tea Companies from India

By the time that Brooke Bond and James Finlay came to develop their new properties in Kenya there was almost a century of experience to draw upon and tea growing had become a mature industry. However, they were to find it impossible to replicate all the circumstances of tea cultivation in India in their new theatre of operation.

In its wild state, *Camellia sinensis* is a forest tree that grows to a height of 30–40 feet. Cultivated in India since the 1830s, tea was grown on large estates, or tea gardens in the jargon of the planters, in closely planted rows. Young bushes were shaped as they grew in order to produce a uniform flat table and leaf canopy some three feet high, which made for a striking visual landscape. Every four years the bushes were ruthlessly pruned down to a height of two feet and the crop was lost for that year while regrowth took place.

A feature of the tea bush is that regular harvesting of the young leaves – the so-called 'flush' of new growth – is possible, and a ten-day plucking cycle was common. Gangs of pluckers from the nearby village consisting of women and youngsters with baskets on their backs worked through the fields gathering the young leaves. The aim was to pluck the apex bud and the next two or three young leaves to achieve the best quality. The harvested leaf was transported to the tea factory for conversion into black tea, following a mechanised version of the traditional Chinese tea-making practice. The first step was to lose a large part (70 percent) of the water content of the green leaf by evaporation in a process called withering. The fresh leaves were laid out on hessian racks (or tats in the jargon) for a number of hours until the desired flaccid condition had been reached, depending on weather conditions and the time of year. Next, the leaves were bruised, without being torn, by mechanical rollers, a process akin to rubbing them between the palms of one's hands, in order to initiate a process known as fermentation; they were then laid out in the fermentation room for a number of hours. The analogy with alcohol suggested by the term 'fermentation' is somewhat misleading, since the chemical reaction comprised the oxidation of the complex organic compounds in the leaves that control taste and colour. A powerful aroma is generated during fermentation and the process is then arrested by transferring the leaf to drying drums into which heated air is fed in order to drive off the remaining moisture. The leaf was now ready for mechanical sorting into its designated grades – pekoe, broken orange pekoe, orange pekoe, fannings and dust – before despatch to the tea auctions or to direct buyers for blending and packaging for the retail trade.

The Brooke Bond story commenced in the 1840s, when Charles Brooke set up a wholesale tea business at Ashton-under-Lyne, at a time when tea was already a popular household drink. His son, Arthur, did likewise opposite the Royal Exchange in Manchester in 1869, naming the business Brooke Bond & Company. The key to its success appears to have been that, instead of selling loose tea, the product was blended for consistency of flavour, packaged and advertised, and then sold to wholesalers under the Brooke Bond label. Brooke visited America and was influenced by the role of advertising there. The company went public in 1892. A buying office was opened in Calcutta in 1900 and the company won the prestigious contract to supply tea to the Royal Durbar in Delhi in 1902; subsequently, it won the contracts to supply the Royal Navy and the army.[17] Brooke Bond commenced its famous 'little red van' distribution to retailers in 1907, and it began a packet distribution business in India, with Brooke Bond India being incorporated in 1912.

Brooke Bond's initial interest in Kenya was as an export market for its Indian tea. Thomas Rutter had been recruited as head of sales in 1905 and, following a misjudgement of the market by the Calcutta office, he took over as sales director there in 1910. Brooke Bond was already exporting tea to Mombasa and in 1914 Rutter visited his sales agents there, and then proceeded up country to Limuru, where he stayed with Caine. Samples of its hand-made tea were sent back to India for evaluation, where presumably they received a similarly favourable report to that of the Imperial Institute. This direct knowledge of the country became significant after the war. A sales office was opened in Mombasa in 1922.

Gerald Brooke became chairman of the company in 1910 (having joined the business straight from school in 1899). In 1920 he visited India, including the newly purchased Taikrong tea estate in Assam. He also went to Ceylon and authorised the purchase of the Udaveria estate. Another estate, Dibru Durrang, was purchased in Assam in the following year. In 1922 an experienced Ceylon tea planter, Malcolm Bell, visited Kenya and made a comprehensive study of the prospects for tea planting there, assembling data on climate and soils and talking to the early experimental farmers. Back in England he first tried to interest the other notable tea merchant, Thomas Lipton, who had purchased a number of estates in Ceylon in the 1880s, but without success. He then approached Gerald Brooke, whom he had possibly met during the latter's recent visit. Brooke took the bait and authorised the purchase of Mabroukie farm in Limuru from Caine and its development as a tea estate. This was undertaken by Bell who, in 1924, recruited a cousin, John Pickford, to come out to Kenya to work for him, and subsequently also Pickford's brother Tim.

Meanwhile, Bell was engaged in much more significant transactions on the other side of the Rift Valley in Kericho, where he became aware of the collapse of the soldier settlement project and that the government had land for sale. In 1924 he purchased 5,000 acres of the Beadoc land at £3.10 an acre. More remarkably, Brooke Bond was able to purchase 6,500 acres of Kipsigis reserve land on a 999-year lease.[18] Bell then recruited his brother-in-law, Murray Clarke, to come out to Kericho from Ceylon as the first general manager to start development there. Six of the independent farms in the district were also purchased, and another four were added in the 1930s and during World War Two.[19] The outcome of all this activity was that Brooke Bond acquired some 14,000 acres of land suitable for tea growing in Kericho, although only 3,220 had been planted up by the time the Tea Restriction Scheme came into force in the mid-1930s and its first tea factory commenced operation in 1934. Having only purchased its first mature tea estates in India and Ceylon a few years earlier, this was a very striking initiative.[20] In making this large commitment to tea development in Kenya,

Brooke Bond was relying initially upon an experienced tea planter from Ceylon, as it had no in-house experience of growing tea, other than drawing on its recently acquired estates. The company subsequently transferred from India experienced sales and engineering staff, and in due course some staff from its estates both there and in Ceylon.

Both Bell and Murray Clarke left Kenya in 1926, leaving Captain Tom Derby as the resident manager responsible for continuing development, with Mills being transferred from India as general manager.[21] Derby had been one of the Beadoc settlers, and then the official liquidator of the project before being taken on by Brooke Bond to open up Kericho estate. John Pickford recalled: "I went to Kericho estate [in 1926] where right up to pretty well the end of planting of some 1,200 acres was all done by teams of oxen... We had at that time something like 500 head of cattle and 6 ploughs going... [Tommy Derby] was bred to the plough and the land and he didn't like machinery of any sort if he could possibly help it, but he was extremely good at clearing land with oxen. We had to fell all the trees by hand. The big thorn trees were very often 90 trees to the acre. Those all had to be cut and burnt and got rid of."[22]

Tea development in Limuru followed a different path following Brooke Bond's purchase of the Mabroukie farm in 1924, and the subsequent purchase of the neighbouring Cainesville property to form an estate of 1,000 acres, of which 425 had been planted to tea by 1938. The neighbouring farmers were keen to grow tea themselves rather than selling out, but they lacked the capital for each to expand sufficiently to support a factory. Instead, Brooke Bond evolved a scheme to purchase harvested green leaf for processing in its own factory, which was completed in 1928. Nine farmers signed up as green leaf suppliers and, in this way, 569 acres were planted by 1933, which had increased to around 800 acres by 1937.[23] The International Tea Restriction Scheme cramped this development for many years.[24]

The concept of a nucleus estate and factory, with outgrowers delivering their harvest to the central factory, became an important model for the World Bank and CDC in the 1960s, notwithstanding its administrative complications. The Limuru scheme is still in operation.

Meantime, the initial development of Mabroukie had its challenges. In 1924, Bell recruited his cousin, John Pickford, to come out from England to work on the estate and Pickford's brother, Tim, joined him in the following year from a banana plantation in Honduras. They had yet to learn to speak Swahili: "The tea seed was imported from India, *Dulia Manipuri* and *Dulia Assam*... I myself transported the tea seed and Mr Howland (owner of Cheymen estate) came down from Kericho and interpreted for us... and we planted the first tea seed in germinating beds and then into nurseries... In those days on Mabroukie there was one house, double-storied, rather decrepit and shaky and Malcolm and Mrs Bell had the downstairs bedroom and my brother and I shared the upstairs room. One had to walk very gingerly about because it shook the whole house and there were various other little difficulties."[25]

James Finlay traces its origins back to the 1750s, when the first James started trading as a cotton merchant in Glasgow and then expanded into increasingly successful manufacturing with exports to Germany, America and the Far East. By the mid-19th century the Muir family were the predominant partners, and it was due to Sir John Muir in the 1870s that the Calcutta branch started taking on tea agency contracts, including advancing working capital to tea estates in order to secure their agency business. The first investments in tea estates were made in the 1880s and, by 1903, James Finlay controlled 74,000 acres of tea in north and south India and Ceylon.

The first African initiative was taken by a Finlay's manager in south India who decided to take his home leave in 1924 by visiting Kenya, where RD Armstrong had relatives working in the Lands Department.[26] While there he was asked to write a report for Glasgow on Kenya's suitability for tea planting. Armstrong came to learn of the collapse of the Beadoc settlement scheme in Kericho and of the liquidator's brief to sell the 25,000 acres that had been alienated for the scheme. He also learned that this was likely to be the only opportunity to acquire a large piece of land in the district,

Brooke Bond having already agreed to purchase 5,000 acres of it. Armstrong visited Kericho to assess the situation and reported favourably to Glasgow, while also stressing the need for a speedy decision. The liquidator (Derby) had refused to grant an option to Armstrong, but revealed that Liptons were also interested in the property. He gave both parties until 31 March (three months) to make a firm offer. Armstrong was able to conduct these discussions without either Lipton or Brooke Bond discovering that he represented James Finlay.[27] The general manager of Armstrong's company was himself due to go on leave in early 1925 and HL Pinches was now instructed to travel via Kenya and report on the situation. He arrived in mid-March and immediately went up to Kericho with Armstrong. His report endorsed the favourable assessment and also urged a quick decision.[28] James Finlay took the plunge and the investment was shared between it and four of the Indian companies under its control. The African Highlands Produce Company was incorporated in September 1925.

William Lee was appointed the first general manager. He was an experienced tea planter, and he brought with him from India a young assistant, Fitz Villiers-Stuart, who subsequently succeeded him on his retirement in 1938. Another young assistant, Hugh Thomas, had expressed interest in the Kenya venture and he was transferred a year later at the end of 1926 (and became the third general manager in due course). Thus Finlay's commenced operations by transferring three trained tea planting managers to Kenya. Lee also recruited one of the Kericho farmers, Hugh Lydford, who sold his property to African Highlands, and who already had African farming experience from Nyasaland. The company purchased two other neighbouring farms (from Homer and Burns), so that its aggregate holding in 1926 amounted to 23,000 acres.

Apart from reporting on the favourable climatic conditions and topography of Kericho and its remarkable deep volcanic soil, both Armstrong and Pinches were mainly exercised about the prospects of building up an adequate labour force for the new estates. The shortage of agricultural workers was a national preoccupation at that time: "There is no getting away from the fact that Kenya has its labour problem... Put concisely, the potential labour is there in the native reserves in the district but the labour is so well off there that outside work is not a necessity for existence." Pinches put this down to the absence of financial incentives to earn money in what had been a subsistence economy, over and above the need to raise enough to pay the hut tax on every dwelling; but he also noted the poor living conditions on settler farms compared with the standards set by the tea industry in India and concluded, "I have little doubt that a large company, working on modern methods, willing to spend money on making their labour force happy, would be able to secure all the labour it required." The company started by employing Kipsigis, who were already living on the land as so-called squatters; but in October 1924 a labour recruitment officer was engaged, to be based in Kisii. A 30-acre tea nursery was ready for the consignment of tea seed from India at the end of 1925. An initial workforce of 500 was assembled, with plans to plant 400 acres of tea in 1926 on land that had long since lost its original rain forest as a result of shifting cultivation by the Kipsigis and now consisted of light bush, bracken and couch grass. It was bad luck that 1926 saw an exceptional 99 inches of rain, instead of the usual 60–70, and only 300 acres of tea were planted, but a wood-and-corrugated-iron house was built for Lee. Anxiety over the labour supply led the company to experiment with mechanical cultivation and to purchase tractors.

A director from Glasgow came out to inspect progress early in 1927 and his detailed report was generally encouraging, especially as regards the growing and climatic conditions.[29] Labour supply was the main concern, and it is interesting that he actually discussed the possibility of importing workers from Ceylon, Italy and elsewhere in Africa, as well as noting the benefit and risk of employing a recruiting agent in the reserves. Meanwhile, the workforce numbers improved steadily, and there was better appreciation of the need of migrant subsistence farmers to return to their homes at harvest time. In 1927, 450 acres were planted, and 683 in the following year, as planting started on three more estates –

Kapsongoi, Kitumbe and Saosa – and preparations were put in hand for the first factory on Chomogonday; it was erected in 1929 and started manufacture in the following year. Unlike Brooke Bond, the company did not have a resident engineer at this stage and it arranged to engage the independent farmer Orchardson to supervise the work.

Concerns over the labour department led to the departure of its head at the end of 1927 and to the appointment of a man from the government service. McInnes, as labour officer in Londiani, had recruited some Kikuyu people for the company in 1926 and "was of considerable assistance in labour matters" over the following year.[30] He put his staff into uniform with a company logo on it and quickly made an impact. He encouraged the visits of chiefs to Kericho and also introduced 'squatter' three-year contracts to people from tribes other than the local Kipsigis. This was deemed a success in helping to stabilise the labour force, but was eventually phased out as more men were prepared to engage in repeating six-month contracts.

It was at this time that African Highlands adopted the symbol of the Kavirondo crane, which also became the trademark when it started to sell its own branded tea in 1930 as Ndege Chai. Brooke Bond had, meantime, adopted the lion symbol for its Simba Chai.

In 1930 James Finlay was offered the purchase of 50 percent of the capital of Buret Tea Company with its 930 acres of planted tea and was strongly tempted. However, Lee eventually advised against the deal on the grounds of the capital expenditure that would be required to bring the property up to its standard (c.£21,000), but also because he still had concerns as to whether Kenya tea quality was up to Indian standards.[31]

1931 was a difficult year. In response to the Depression, Glasgow imposed cuts in the wages and salaries of the supervisory staff, and postponed the construction of the second factory on Kitumbe. Talks were taking place on an international agreement to restrict tea exports and Lee was warned that development work might have to be curtailed. Meanwhile there was intense competition for sales in the local market between Simba Chai, Ndege tea and the Buret Tea Company. To cap it all, the Kenya government chose this moment to impose an excise duty on manufactured tea, without consultation with the tea companies, and competition was such that they absorbed the impost rather than pass the cost onto consumers. This development led to the formation of the Kenya Tea Growers Association (KTGA) as a body to represent the interests of the industry to government. An independent producer was the first chairman and Thomas was appointed secretary.

In the event, development continued on four new estates: Kapsongoi, Kitumbe, Saosa and Changana. The Kitumbe factory was erected in 1933 and that at Saosa in 1935. By 1934, 5,042 acres of tea had been planted over six estates, and with three factories to service them once Saosa was completed. As a result of the International Tea Restriction Scheme, new planting ceased after 1933 and was not resumed for 15 years.

Kericho's only link with the railhead at Lumbwa was 22 miles away along a winding dirt road, over which all incoming equipment and materials had to be trucked, and the export traffic of finished tea expedited. In the rainy season the road became virtually impassable: in 1930 arrangements had to be made to transport the factory output by porters to Lumbwa.[32] The tea companies had lobbied the Public Works Department for years with little result until eventually Brooke Bond and African Highlands made a joint approach that they should assume responsibility for maintaining the road, with whatever funding the government could provide. A visit by the Governor, Sir Joseph Byrne, to Kericho in 1932 provided a timely opportunity for some further persuasion. He promised support and the two general managers were summoned to Nairobi to attend a meeting, where they encountered the usual problems relating to creating a precedent and unbudgeted expenditure. Byrne lost patience and said, "Well gentlemen, if you want my view I think that it's high time that we all had a pink gin." And the matter was settled. Four years later, in 1936, funds were found to tarmac the road from Lumbwa to Kericho, the first stretch of highway tarmac in the colony.[33]

When land was allocated for settlement in Nandi district before and just after World War One the focus of interest was mainly on coffee growing. However, the conditions proved unfavourable due to coffee berry disease and most farmers faced financial difficulties. With tea growing under way in Kericho by the mid-1920s, this crop now exerted strong appeal. The provincial administration and Agriculture Department were very supportive of the claims of the settlers as they confronted the provisions of Kenya's adherence to the International Tea Agreement.[34] The Agreement was designed to arrest new tea development, and especially new entrants, so that the acreage allocation to Kenya in 1933 went to already established growers where they had yet to establish economically viable units. Nandi therefore lost out completely in the first allocation period, but a big effort was made to secure allocations in the renewal period starting in 1938. Only EG Myers from Australia had managed to establish a small estate of 78 acres, where tea was hand-made and sold locally. In the event, the much-scaled-down allocations were 625 acres for eight farms to the Nandi Planters Association, out of 16 bids, and of 330 acres for ten farms to Kaimosi farmers.

In the event, the allocation of 402 acres to Mokong (Myers) was not implemented. The capital costs of development and the waiting period to maturity were beyond most pockets. However, it meant that Nandi/Kaimosi was the one tea-growing area in the White Highlands that was very much open to new investment after World War Two.

Notes to Chapter 1

[1] There are several accounts of the history of tea, including *Tea for the British*, Denys Forrest, Chatto & Windus, 1973; *Tea*, Roy Moxham, Constable, 2003; and *Green Gold*, Alan & Iris MacFarlane, Ebury, 2003.

[2] *Land Reform in the Kikuyu Country*, MPK Sorrenson, OUP, 1967.

[3] MP/EDK/1/2: Note on the History of Tea Growing in Kenya, HO Thomas. MP/EDK/1/1, J Wilson memoir. MP/IM/1//11: Joyce Pickford recorded an interview in the 1970s with her husband Tim and his brother John about the early days in Limuru and Kericho.

[4] Director of Agriculture to Director of the Imperial Institute. See note below, 2.12.1908.

[5] MP/EDK/1/2/d, *Report on Tea from the East Africa Protectorate*, WR Dunstan, 26 March 1909. This report was reprinted in a pamphlet about the agricultural potential of the district published in 1916 by the Limuru Farmers' Association, which refers to the letter from the Director of Agriculture above. It was reprinted in *East Africa: Its History*, People, Commerce, Industry & Resources, ed. F Holderness Gale, Foreign & Colonial Compiling and Publishing Company, 1908.

[6] MP/EDK/1/2/a, Letter from IQ Orchardson to HO Thomas, 21 June 1935.

[7] MP/1M/1/1. Wilson was later a manager of the Buret Tea Company.

[8] Orchardson, op. cit.

[9] MP/ITA/1/1, Memorandum by director of agriculture, June 1938.

[10] MP/Lib/27, *The Kipsigis*, IA Orchardson, The East African Literature Bureau, Nairobi, 1961. The place also has other names and my informant in 2003 told me that the name was Toiobisho Hill. MP/1M/3/3, Kotir. The settlers called it Londiani Hill.

[11] *Some Aspects of Kipsigis History before 1914*, SC Langat, East African Publishing House, Nairobi, 1969. See also MP/1M/3/3, Kotir.

[12] *Report of the Kenya Land Commission*, Cmd 4556, 1934. Vol lll Evidence. Letter to provincial commissioner Nyanza, 23 December 1925. Hemsted was a member of the Carter Commission.

[13] Ibid.

[14] Memorandum Native Rights to Land, 23 September 1932, Carter Report, p2458.

[15] Carter Report, p2467.

[16] The subsequent sale of the Beadoc land to Brooke Bond and James Finlay realised £90,000, which was sufficient to pay off all creditors and to distribute £800 each to the remaining 36 members.

[17] MP/Library/1, *Brooke Bond – A Hundred Years*, David Wainwright, published by the company in 1970. The book is a celebration, rather than a corporate history, with much emphasis on the Brooke family. There is very little detail on the development of the tea estates in Kenya. My father was posted to Kenya to join the sales office in Mombasa in 1925, before being transferred up to Kericho as an assistant to develop the Cheboswa estate, where I was born.

[18] I have been unsuccessful in sighting the original leases, LR5467 and LR3939. They are mentioned in a letter of 7 April 1949 from the Member for African Affairs to the provincial commissioner of Nyanza, on which the district commissioner commented on 20 April, "The leasing of Kerenga-Chebown is still clothed in mystery and I am by no means happy in my own mind that the lease was effected with the consent of the Kipsigis... I have been unable to find any record of the Local Native Council and Local Land Board's consent, and the Kipsigis are insistent that they never approved the transaction." MP/BB/1/6. There seems to be a mistake in the 7 April letter, which states that the leases were granted in 1920, whereas Bell only visited Kenya in 1922 and contacted Brooke Bond thereafter. KNA/LND/9/1/7.

[19] Two farms on what became Kimugu estate, belonging to the Orchardson brothers and to Grice; Howland on Chemen (where he became the manager); Fletcher on Cheboswa; the mission property on Chagaik; Le Breton on Chepsir; and the Gailey Syndicate on Kitoi. In the 1930s, Caddick's farm at Chelimo; Buchanan's at Kaptien and Sneyd's at Kimari were added. MP/EDK/1/1, Wilson memoir.

[20] There is a puzzle over the timing of the Brooke Bond purchases and some conflicting evidence, as noted. The most plausible sequence appears to be as follows. The fact that Brooke Bond had just purchased tea estates in India and Ceylon means that the Board was aware of the value of growing tea as well as selling it, and there was the added merit of getting an advantage over competitors in East Africa by getting behind the tariff barrier. What is unclear is precisely when Bell made his visit to Kenya and when he presented his report to Gerald Brooke. The steps in the sequence, with plausible dates in brackets, were as follows:
Visit to Kenya from Ceylon [1922]; report presented first to Lipton and then to Brooke [1922]; approval to invest and Bell's return to Kenya [1922]; purchase of Mabroukie [1923]; purchase of Beadoc land [1924]; purchase of lease in Kipsigis reserve [1924]; Bell's return to UK to recruit Pickford brothers and Murray Clarke [1924]; development commences on Mabroukie [1924]; development commences in Kericho [1925].

The Brooke Bond history makes no mention of the lease of the Kipsigis land, although it became its first estate. There is another possible explanation of the purchase arrangement: in 1920 the Kenya Government approved two large transactions in Kericho. One was the 25,000-acre grant for the Beadoc soldier settlement scheme, and the other was a 7,000-acre grant to the Stevens Syndicate. They are recorded in the White Paper 'Crown Grants: Kenya', Cmd 2747,

1926. Perhaps Bell subsequently did a deal with the Stevens Syndicate.

[21] The removal of both Bell and Murray Clarke is unexplained, but leaves a distinct impression of a serious falling out with the chairman, Brooke, in London. It is notable that Bell is airbrushed out of the corporate history, despite his key role in bringing the company to Kenya.

[22] MP/IM/I/II, Joyce Pickford's recorded memoirs of John and Tim Pickford, August 1971.

[23] Brooke Bond records show nine green leaf suppliers in 1937 and during the war: Buxton, 70 acres on Kabuku farm; Charteris, 75 acres on Honinger farm; Delaney, 60 acres on Mount Blair; Fowler, 70 acres on Nyara farm; Franklin, 70 acres on Satimi farm; Limuru Tea Company, 187 acres on N'Deri estate; Morson, 80 acres on Gwalia farm; Simpson, 130 acres on Tigoni estate; and Tunstall on Gathiga farm. The acreage figures are from a memorandum by the Director of Agriculture, dated June 1938, reviewing representations by settlers to plant tea under the International Tea Restriction Scheme. MP/ITA/I/I.

[24] The Limuru farms applied for an aggregate of 310 acres of tea for the period 1938–43, but were only allowed 197 acres between them. Statement by Director of Agriculture to Nairobi Chamber of Commerce, 14 July 1939. MP/ITA/I/I.

[25] Pickford memoirs, op. cit.

[26] MP/Lib/2, *James Finlay & Company 1750-1950*, published privately by the company in 1951. MP/Lib/3, *A Brief History of the African Highlands Produce Company*. This account of the development of the Kericho estates was written by Hugh Thomas in 1959. He was my stepfather.

[27] Report by RD Armstrong on Beadoc land, 1 January 1925. This is included as Appendix 2 of the *Brief History*.

[28] Report by HL Pinches on Beadoc land, 3 April 1925. This is included as Appendix 3 of the *Brief History* The text had been cabled to Glasgow on 19 March.

[29] Report by R Langford James on his visit to Kericho, February 1927. Appendix 4 of the *Brief History*.

[30] Thomas, p.20.

[31] MP/AH/I, Lee to Finlay head office, 23 October 1930.

[32] MP/BB/2, Estates Circular 19 May 1930.

[33] Thomas, p.38.

[34] KNA/PC/RVP file 6A./II/IA, Memorandum, '*Tea Planting in Nandi District 1937*'.

2

TOWARDS A SETTLED LABOUR FORCE

The biggest challenge facing African Highlands and the Kenya Tea Company in 1925 was to recruit enough labourers to undertake bush clearing in order to prepare the land for a tea nursery and, subsequently, the plantation. As he left India for Kenya to initiate development for African Highlands, Lee was informed that enough tea seed for a 30-acre nursery had been ordered and that he had approximately three months to prepare for its arrival. The initial workforce comprised Kipsigis men already living on the land that had been sold to the tea companies and who were now designated resident labourers, or squatters in the local settler parlance. They proved effective in the bush-clearing phase and the nursery was ready in time. But it was immediately apparent that this was no solution to the problem of a stable workforce as it was not customary for men to engage in cultivation, which was traditionally women's work. Lee decided to engage a labour recruiter, whose task was to hire migrant workers from further afield in Nyanza Province. The Kenya Tea Company, meanwhile, under the management of the ex-Beadoc settler Derby, built up a local workforce and subsequently left recruitment to the manager of each estate as it was developed. But the requirement for labourers was less, as Derby was an enthusiast for ox-ploughing, and then tractors.[1]

The initial 100 labourers engaged by African Highlands in April had increased to 500 by the turn of the year and reached 1,032 by June 1926. There was then an alarming reduction during the rest of the year to 444, as workers returned to their own farms for the harvest. However, numbers built up again and reached 2,683 by June 1927, of whom approximately half were juveniles. Again there was a dip, but the workforce was restored to 2,597 by April 1928. The basis of employment was the 'ticket' – which entailed an undertaking to provide 30 days' work over a 42-day period, and the norm was to sign up for three tickets at a time. The hope was that a worker would then return after leave for another three-ticket assignment. In 1927, African Highlands started encouraging workers to sign up for six tickets at a time, and this soon became standard. The men were housed – four to a hut – and provided with rations and a blanket for every six tickets. African Highlands had a significant number of Kipsigis squatters on its property, where there was now a legal obligation to undertake paid work in return for continuing occupation. In 1928 it was decided to offer similar terms – under a three-year agreement – to workers from outside the district, who would also be encouraged to bring their families with them. In this way it was hoped that many would settle on a permanent basis. The policy was successful, but over time the same outcome was achieved under the six-ticket contract as family housing came to be provided, and the three-year agreements lapsed save for those for the resident Kipsigis.

During these difficult early years, African Highlands more than once considered the possibility of importing indentured labour from outside the country, reflecting tea industry experience in Assam and Ceylon, but it received no encouragement from the colonial administration. However, its Indian experience strongly reinforced the desirability of family work units on the estates, so that women and children could be employed, especially for plucking. In the Ceylon highlands and in Assam, there were village communities attached to the tea estates where all able-bodied residents could find work. By May 1927 half the African Highlands workforce of 2,350 comprised juveniles. In 1928, Lee decided to hire a member of the Labour Department, who had been particularly helpful to the company, to manage recruitment in the reserves and McInnes made an impact. Offices were set up throughout Nyanza Province; McInnes increased the scope of the squatter agreements and promoted the six-ticket employment contract.

The Kenya Tea Company became increasingly aware that its recruitment arrangements were not as effective as its rival's and considered establishing its own labour department. This would have resulted in overt competition around the province and a likely increase in costs. The two companies began discussions over establishing a joint operation and this was eventually accomplished at the end of 1938, with McInnes in charge.

The Juvenile Labour Controversy

In the early colonial period the Kenya government came under strong pressure to motivate the native population to engage in wage employment on the newly establishing farms and plantations, as well as in the public sector. To this end, legislation was introduced to levy hut tax and a poll tax, which had the effect of forcing workers to leave their subsistence cultivation to seek wage employment. Officials were also expected to use their persuasive powers. The situation was monitored closely by churchmen and others in England, who were concerned that a line might be crossed between persuasion and coercion and also about what the social consequences of migrant labour might mean in terms of disruption of community life. The creation of a plantation society in Kericho brought these issues into sharp relief for the tea companies. From their Indian experience, the idea had become axiomatic that plucking was women's and children's work, and a mythology grew up: "Children are particularly suitable for light agricultural employment where their small nimble fingers have a definite economic value."[2] However, few paused to reflect that the new Kericho estates were being established on substantially empty land without surrounding villages, and that labour recruitment from all over Nyanza Province meant that young people would be leaving their homes for prolonged periods. The urgent need for an expanding labour force had blunted sensibilities.

Administration officers themselves were puzzled as to what attitude to adopt towards this aspect of the employment scene: for example, the provincial commissioner's conference in May 1927 debated the matter inconclusively, only to resolve "that Government should declare a definite policy in regard to juvenile labour."[3] There was little disposition to challenge the existence of juvenile labour, and the discussion concentrated instead on its application and whether places and kinds of permitted employment should be defined. The apologists argued that the children were only employed on light work, and that their early introduction to the discipline of manual labour could only be to their good and to the advantage of the community. Much was made of the benefits of contact with civilisation. As regards recruitment, there was a tendency to argue, as did the senior commissioner for Nyanza in 1926, that "children cannot go away without their parents' sanction. If their guardians do give leave, I don't see how they can be prevented from going." It was merely a domestic matter and beyond the jurisdiction of the outside authorities.[4]

But for those observing the working of the system, a rather different picture emerged, both of the conditions of employment and the circumstances in which recruitment took place. It became apparent that children were often persuaded to leave home by native recruiters without obtaining parental consent. As a labour officer wrote of one instance after interviewing the parents of six children who had been picked up as far away as Nairobi: "I am satisfied that the parents not only did not consent but were kept in ignorance by the recruiter lest they intervene."[5] Under the Masters and Servants Ordinance the only provision for children was that under the age of 16 they could not sign work contracts. Yet children from the age of seven upwards were leaving the reserves and being employed on verbal contracts.[6] The missions became concerned and formally asked the government to forbid the employment of juveniles in towns; yet one mission was discovered to be sending children to work on a nearby coffee farm "to acquire practical knowledge in using their hands" and was formally reprimanded by the senior native commissioner.

In 1927 the Chief Native Commissioner obtained a copy of Southern Rhodesia's Native Juvenile Employment Act of 1926. The principle of the Act was that juveniles in employment should be specially registered, providing there was no objection from the family to their going out to work. He recommended that an ordinance on similar lines should be applied to Kenya. But, in the event, the first legislative action was to follow the lead of the International Labour Office and the Employment of Women, Young Persons and Children Ordinance of 1933 enacted International Labour Organization conventions that had been recommended to the Kenya government twelve years previously by the then colonial secretary, Winston Churchill. This ordinance provided that no child under 12 years of age was to be employed in industrial undertakings or on ships. By implication this met some of the missionary misgivings about conditions of work in towns, but said nothing about agricultural employment or recruiting methods.[7]

In 1936 a new Employment of Servants Ordinance was considered, which included clauses on juvenile employment. The labour officer for Nyanza summarised the views of the administration at the time: "Whilst it has been generally agreed that some form of control is necessary, no attempt has yet been made to enforce any system of control. The consensus of opinion seems to be that it is beneficial for juveniles to go out to work after they have attained, say, the age 10, but that it is understandable that under the age of 12 they should not go too far from their homes." The Kenya Tea Growers Association was consulted when the bill was in select committee and its view was that juveniles should be registered under the Registration Ordinance (so that they would have to carry employment passes) and it was not opposed to the requirement for parental consent. When the Ordinance was enacted in 1938 it provided that no children under ten years of age could enter into a contract of service and no children could be recruited without a certificate showing parental consent (as attested by the district officer). Such contracts were subject to penal sanctions. The provisions of the Ordinance did not apply to children employed by the day when accompanied by adult relatives.

The juvenile sections of the Ordinance were not well received by the tea industry. It complained that the arrangements for parental consent would prevent any children coming out to work. Missionary critics and their supporters in England were startled that juveniles could be recruited for employment from the age of ten. In June 1938 a question was tabled in the House of Commons to the colonial secretary, Malcolm Macdonald: "What is the nature of employment of children at 10 years of age?" He flannelled his answer: "It is really employment in the plantations. I understand that the principal kind of employment is picking foreign bodies off the tea trees. It is all employment of a light kind."

The Kenya Government decided to set up an investigating committee before applying the juvenile sections of the Ordinance. In its report, the committee attempted to estimate the size of the juvenile labour force in the colony and estimated that, at any one time, there were some 14,000 children in employment and that the Kericho tea estates were the largest employers, with some 5,500 children.[8] However, due to the high turnover rate, it was further estimated that during a year some 36,000 juveniles would have been in employment. The tea estates were praised for the conditions they provided, and especially for the start being made with schooling. The committee concluded that, under such conditions, juvenile employment could do no harm and it endorsed the general principle: "In Kenya, as elsewhere, the general principle of employment of juveniles has already been accepted. Criticism has generally been aimed at the terms and conditions of employment rather than the principle. Under proper safeguards and stringent regulation we do not consider it to be harmful." The economic importance of juvenile labour at this time, which constituted the tea plucker element of the workers, is illustrated by the following figures from Kerenga estate in Kericho:[9]

	1934	1937	1939
Men	280	733	655
Totos	391	397	491
Total	**671**	**1,130**	**1,146**
Juvenile %	58%	35%	43%

The KTGA had met the committee and put forward several suggestions that were accepted in the report: that penal sanctions for juveniles be abolished and civil contracts only be applied; that the minimum age be raised to 12 years; and that the attestation procedure be simplified.[10] The government accepted the committee's report and the Ordinance was appropriately amended in 1939.

The report marked a distinct advance in local opinion on juvenile employment: the willingness of juveniles to go out to work needed to be controlled if the economic advantages were not to be more than offset by social evils. A minimum wage and parental consent were obvious first steps. The committee also drew attention to the connection between juvenile employment and the lack of education facilities in the reserves, and it clearly envisaged that its attraction would diminish as educational opportunities became more available. The report went so far as to recommend "that meantime there is an obligation on employers who employ juveniles in any appreciable numbers to provide educational facilities, and upon the Government to assist, particularly by inspection and advice."

The larger tea companies in Kericho were, on the whole, well in step with the committee's views: penal sanctions had never been used and 12 years was their minimum age for employment. A start had been made with schooling, although it only embraced a small fraction of the juveniles in their employment and the standard was poor. It was a weakness of the committee's report that it did not make more definite recommendations as to the obligations of employers, which would have stimulated improvements. However, there was still a large gap between enlightened opinion in Kenya and the missionary critics and their pressure group in England, who were disappointed by the report and the laissez-faire attitude of the government: "The definite evil, wrote Archdeacon Owen to the *Manchester Guardian*, especially in the towns, but also on the sisal, tea and other estates, is that we should bring into existence systems of labour which break up the child's connection with the home to earn wages at an age when it is most harmful to do so."[11] Writing also to the Colonial Office of his deep disappointment at the failure of the committee to consider the implications of juvenile employment on home life and tribal society, Owen concluded: "We are rushing Africans, children and adults, far too swiftly into our modern industrialism. We have an opportunity to put on the brake where children are concerned. I trust the importunity of industrialists will not be allowed to win the day."[12] It did.

The safeguards written into the amended Employment of Servants Ordinance were never implemented. The amended Ordinance was passed in June 1939, but it was provided that it was not to come into force until declared by the governor. Three months later, war was declared and the officials responsible for administering native registration asked that the juvenile sections of the Ordinance be deferred because all available staff were required for the enforcement of the Ordinance to adults. This was agreed to.[13] Later, in 1943, the Colonial Office tried to get the juvenile sections applied, but by that time there was an acute labour shortage in the colony resulting from the demands of the military. Conscription had been introduced for essential industries. Implementation of the Ordinance would have entailed the repatriation of all juveniles so that the certificates of permission could be obtained. This would have caused great dislocation on the estates and it could have been anticipated that many juveniles would not return. The Labour Commissioner met the KTGA and together they drafted a strong statement.[14] No more was heard of the matter until 1949, with the appearance of a new Labour Ordinance.

By this time the tea industry was strongly against juvenile workers having to obtain a permissive certificate from their parents. It feared that under the post-war conditions of prosperity and increased educational opportunities in the reserves, the flow of workers would be reduced to a trickle if parents had a deciding voice in the matter. The companies were not prepared to envisage the end of juvenile employment since, quite apart from cost considerations, labour was by no means plentiful. Instead, a compromise was proposed that the minimum age be raised to 14 years, above which there would be no restrictions on recruitment.[15] However, the Labour Commissioner (Hyde-Clark) rejected the proposal, arguing that until there were sufficient schools to provide compulsory education for these children up to the age of 14 "I think we should be doing them a disservice by preventing them entering employment."[16] He insisted on applying the provisions requiring parental consent, although the procedure was simplified, and they came into force at the beginning of 1950. For the first time the views of parents and chiefs were given the support of law. The effect was dramatic: the flow of recruited juveniles fell by 95 percent. This resulted in a crisis in Kericho and the deputy labour commissioner was sent to report on the situation.[17]

He found that the introduction of the certificate was a popular innovation since, for the first time, parents had the backing of the law against the influence of the recruiting officers, who were deemed to be enticing away their children. The chiefs were also against juveniles going out to work on several grounds. It was asserted that in many cases children were already attending school and fees had been paid. It was said that the loss of parental control weakened tribal unity and traditions, and that the children came under bad influences while away. There was a loss of domestic help and children rarely came home with any money. The commissioner noted that a serious weakness of the Ordinance was being exploited, in that in only applied to juveniles who had been formally recruited. This meant that if a juvenile showed up seeking employment of his/her own volition there was no obligation then to obtain parental consent, nor did the minimum wage apply in such instances.

A meeting was convened between the KTGA and the administration at the beginning of May 1950 to resolve the crisis.[18] It was common ground that the flow of juvenile labour must be resumed, while meeting as far as possible the wishes of parents and chiefs. The introduction of the certificate had shown that if parents were asked whether their children could leave home to go to work they usually refused. The solution adopted therefore was to place the initiative on parents to raise an objection for their children to leave. But no restriction was placed on recruiters' activities, beyond an obligation to provide the chief with a list of juveniles hired. Objections were to be heard by the labour superintendent (only for African Highlands and the Kenya Tea Company) or a district officer before recruits left for Kericho. The tea companies also gave an undertaking that if a parent subsequently went to a district commissioner with good reasons why a juvenile should be repatriated then they would do so at their own expense.[19]

These arrangements proved satisfactory to the tea companies in that they were able to resume juvenile recruitment. The new arrangement applied only to the tea companies; elsewhere in the country, a parental certificate was still required. However, as the post-war employment situation improved, the reliance on juvenile employment eased, and this decline was hastened by improvements in education facilities in the reserves and by its unpopularity with parents. By the mid-1950s the change of tone was striking: "At a guess 30 percent married men to bachelors might be taken as an average figure. Women are not always willing to work, but when they will do so suitable jobs can be found, such as light weeding and plucking. From the employer's point of view children are a liability which he has to accept. On most of the estates they are educated, receive free medical attention and often assistance with extra food."[20] An industry practice inherited from India and Ceylon gradually faded away in the face of improvements in education and a deepening of the labour market.

Workers or Farmers?

The KTGA had been set up in 1931 in response to the shock of an unanticipated excise tax on tea, when the tea companies recognised that they had a common interest in establishing a body that could make representations to the Kenya government. The secretariat, for its part, welcomed an official interlocutor for dealing with the tea industry on the issue of employment of juveniles, and again over the application of the International Tea Restriction Scheme. The tea companies themselves soon found that the forum provided a valuable opportunity to discuss the perennial problems of labour supply and the proceedings of the KTGA meetings show the responses of the industry to the evolving scene.

Wartime conditions had created considerable problems for the tea industry and other large employers, since they were competing with the military for recruits. In 1942 an official committee to examine the problem concluded that the manpower existed in the reserves to alleviate the shortage, but that legal compulsion was necessary to obtain workers because voluntary recruitment, even with official assistance, was not delivering the requisite numbers, nor would raising wages solve the problem: "It has been stated, and we endorse this, that as a rule, a native earning 9shs will not do more work if offered 10shs or 11shs. Some are happier either sitting at home in their reserves or on a farm doing an easy task for a comparatively low wage. In such cases the conditions of life are a more important factor than the actual cash remuneration."[21]

Archdeacon Owen, a member of the committee, dissented; he thought the wage level was too low and was not an economic wage for a family. However, he did not call for an increase, but rather that the hidden subsidy from the family holding should be replaced by an open one from the government. In the event, the problem was alleviated by a cut in military recruitment, especially in Nyanza Province. Nevertheless, in 1943 the tea industry was compelled to release 3,500 workers for the military, and also to cut its standard ration of maize by a quarter.[22] On top of this, the companies had to contend with the Colonial Office's attempt to introduce the new provisions on juvenile labour.

When James Finlay first came to Kenya in 1925, the notion of importing labour to develop the tea estates had been raised with the government, in the knowledge that the tea estates in Assam and Ceylon had been developed with imported workers, who had then settled in villages on the estates. It had been rejected at the time, but in the light of wartime labour shortages the idea was revived, and this time the government was sympathetic to a proposal in 1943 to recruit contract workers from the Belgian Protectorate of Ruanda. The Hutu people there were an impoverished underclass and already many of them sought work in western Uganda. A year's negotiations matured with a first batch of workers coming to Changana estate in 1944. The experiment was deemed a success and was expanded until the Hutu accounted for about 10 percent of the workforce. The arrangement continued until the late 1950s, shortly before the independence of the Congo.[23]

In 1944 the tea companies made a major effort to agree common pay levels and ancillary benefits. Agreement was reached on maximum wage rates for adults and juveniles, overtime rates, food rations (maize meal, sugar, beans, salt), a blanket issue and defining the daily task before overtime payments. The most contentious issue was the earning capacity of pluckers on overtime in relation to the basic wage when there was a heavy flush on the estates. The KTGA went on to debate establishing a joint recruitment agency for the industry, building on the African Highlands/Brooke Bond one, but they were not keen to share their operation.[24] The issue was still an aspiration a year later.

The late 1940s and the 1950s were challenging times for employers: there was a general desire for a more stable labour force to improve skills and productivity and to mitigate the costs of a regime of 100 percent annual labour turnover; there was recognition that wage levels did not constitute an independent living wage and that they were predicated on a family home in reserves as back-up; there was apprehension and reluctance to face up to the capital costs of married-quarters villages and the implications of a leap in wage costs; and there was an awareness within the government administration that colonial practice should reflect developments in Britain on the organisation of labour and social security.

With the cost of living rising after the war, and labour recruitment still difficult, the question posed by Archdeacon Owen back in 1942 had renewed force, and the tea companies themselves now acknowledged that the wage level was a material factor in recruitment. A KTGA meeting in 1947 considered proposals for revised pay scales, where the chairman "drew attention to the general unrest in the colony brought about by inadequate wage levels in relation to the present high cost of living," and tabled proposals for new wage scales.[25] These laid down maximum rates of pay that were 25 percent above those approved three years earlier, in 1944. The issue remained topical and, in 1949, the Association revised the general level of wages again, raising them by 20 percent: "The chief point emerging from the discussion was that most members felt that in recent years the African's rate of pay had been overlooked by the Tea Industry and now was the time to put the matter right. The proposed increased rates of pay would be more in line with wages now being paid in other districts and other industries and would probably prevent labour leaving the Kericho area."[26] The difficulty in reaching a common standard on pay was that the KTGA embraced several relatively small employers who were very reluctant to follow the standard advocated by the Kenya Tea Company and African Highlands, and discussions could be very contentious.

Despite these steps to improve the economic attraction of wage employment in Kericho, the reality was still that small-farm income (and security) was more attractive for those who had the choice. This led to some frustration in the tea companies that the colonial administration could be doing more to tip the balance in their favour, especially as the companies were now in a position once more to expand the acreage under tea, as is illustrated by a KTGA meeting in 1950: "The land is available, so is the capital, and the only factor which prevents an early expansion on very consid-erable lines is the shortage of labour."[27] With the Commissioner for Labour (Carpenter) and other officials present, two proposals for alleviating the labour shortage were aired. The first was a bid to establish a priority on labour recruitment for the tea industry: that it should be given priority recruitment rights in Nyanza Province and that other large employers should be directed to other parts of the colony. The second went to the heart of the tension between sustainable small-scale farming and the demands of the wage economy: that too much maize was being grown by small farmers as a cash crop at the expense of labour supply. "Would it not be a sounder policy to restrict this crop to more reasonable dimensions, in the interests of the future productivity of the soil, and encourage some of the inhabitants to go out to seek work and earn money outside?... Whilst it is not contended that adult labour is actively discouraged from going out to work, it is definitely suggested that it is actively encouraged to stay in the reserves to grow maize, and work on other projects therein, and this amounts to very much the same thing. It is now urged that, in the interests of the general production of the Colony, there should be active encouragement on the part of Government to get a higher proportion of labour out of the reserves into the various industries that need it."[28]

Having worked off its resentments, the meeting went on to discuss labour recruitment, and the industry was congratulated on its procedures and conditions of service. The head of African Highlands, Villiers-Stuart, then made a momentous intervention. His welfare officer had been doing careful research on the household budgets of bachelors and families and Walter Wilkinson had worked out that the maximum annual earnings of a worker on the basic wage plus long-service bonus were Shs240 per year, whereas the cash needs of a bachelor were put at Shs216 (before any saving for bride price, i.e. money paid by a man to his future wife's family); those of a married couple with one child were put at Shs390 per year; and the cash needs of a couple with two children were put at Shs443 per year.[29] Armed with this information, Villiers-Stuart said that he considered workers to be underpaid; indeed, he thought that although standards and expectations had risen since the 1930s, employees were worse off than in 1937 and, of course, they did not share in the profits of the industry. He concluded by saying that a 'just rate' for employees was Shs1 a day – implying Shs30 a work ticket (and Shs327 per year) and a jump of two-thirds compared with the present basic wage.

The concept of the 'living wage' independent of income from a family farm was now firmly on the table. The KTGA submitted a paper to the Board of Agriculture in December 1950 which commenced "It is considered that the great majority of agricultural labour is not paid a living wage."[30] The paper summarised the cost of living research. It noted the costs and inefficiencies of the high turnover from a migrant labour force and (echoing Archdeacon Owen) observed, "In effect, the reserves are subsidising employed labour in the settled areas. This is basically wrong and will eventually lead to discontent and possibly trouble. It also prevents the establishment of permanent and efficient labour forces." At the meeting itself, the KTGA chairman, Angus Kerr, brought the points together succinctly: it was important to pay a living wage to reduce dependence on the reserves and to provide conditions for establishing a permanent labour force. "The proposal for a living wage has nothing to do with efficiency; a living wage should be paid and the question of efficiency would come afterwards."

This striking initiative, led by African Highlands, did not command universal support within the KTGA, and a compromise had to be forged. The outcome of the heated debate was to agree on a starting rate of Shs24 a ticket (compared with Villiers-Stuart's assertion that it should be Shs30). This still entailed an increase of one-third in the basic wage, and it indicated how dependent tea workers still were on support from home.[31] Reservations about this development were not only on grounds of the sharp jump in labour costs. Floyer was one of the original farmer settlers in Kericho before joining African Highlands, and he was now manager of Kaimosi estate in Nandi. He wrote a thoughtful paper in 1950, arguing that to raise wages for a migrant worker farmer could have the perverse effect of reinforcing his position in the reserve: "We will be subsidising the Reserves, helping in their overcrowding and not, as the notes on the resolution suggest, the Reserves subsidising employed labour." The real challenge, he argued, was to establish a resident settled labour force on the tea estates "to de-tribalise the African". He acknowledged the opportunity cost of giving up tea land and the difficulty of making estimates but, "1 think that a deliberate attempt should be made to settle labour on the land owned by the growers... this will require a good deal of organisation, not so much in the nature of films and football as in the nature of Agricultural Welfare."[32]

Floyer's memorandum was an early recognition that the economic development of Kenya would require the permanent movement of people out of the rural areas and into employment in modern sectors of the economy, and for them to be self-sufficient there. At this time there was general concern that the African land units – the reserves – would not be able to accommodate a growing population that was mainly engaged in traditional practices, and that something should be done about it. The Governor of Kenya, Sir Philip Mitchell, called for a Royal Commission to address these issues in November 1951 and one was duly appointed the following October, just as unrest in the Kikuyu districts exploded into violence and the Mau Mau rebellion, so that a state of emergency was declared in October 1952. Notwithstanding these dramatic developments, and to a significant extent also in response to them, the administration took the initiative to launch Kenya onto a fresh trajectory. This embraced land reform to consolidate scattered holdings and to issue land titles, improved farming with cash crops and an enquiry into African wages led by the labour commissioner, whose central preoccupation was to establish a viable basis for a workforce divorced from rural roots and subsidy. These initiatives were matured even while the Royal Commission was sitting and before it eventually reported in 1955.[33]

In May 1952 the government asked the tea industry whether a resident labour force was achievable.[34] It was concerned about the future labour requirements of the tea industry as acreage expanded and wondered whether the existing short-term migrant labour pattern would be adequate. Would it be a practical proposition to create a resident labour force living permanently on the estates, as was the case in Ceylon and parts of India? And could more women be employed as pluckers and in the factories? A special meeting was convened to consider the response, at which there was concern as to whether the government was trying to foist a social security system onto the industry to provide villages for retired people and whole families and to give up its land rights. "It was agreed that the

Tea Industry cannot shoulder the burden of a large excess population of non-working elders and children. To create a resident labour force is in the view of the Industry, a working labour force."[35] At a follow-up meeting, the head of Brooke Bond "said he was attracted by the idea of a resident labour force and would be prepared to experiment with such a scheme. He suggested providing each family with a house and half an acre of land", without title.[36]

In its formal response to the Carpenter Committee in June 1953, the tea industry made some significant commitments: to pay its permanent workers a self-sufficient family wage; to provide family housing with a vegetable allotment; and to provide primary schools. It accepted an aim to move to cash wages without rations. It took a firm view that land and housing for retired workers was a public responsibility, but it was in favour of legislation to introduce pension schemes for workers.[37] The KTGA's assertion that it was already providing a living wage was undermined by the fact that wage rates were still at the level established in 1950 and below what Villiers-Stuart had then argued were necessary to meet that test. However, a further one-third increase in the wage level was agreed at the end of 1954, after the report was published.[38] By the mid-1950s the tea industry was on a path to stabilising its workforce, but there was still a long way to go. Only about 10 percent of the workforce could be regarded as permanent at that time, and there was negligible employment of women. The vast majority of workers still retained a home in the reserves, to which they returned at regular intervals. Out of an industry total of 27,847 workers in 1955, 83 percent came from the neighbouring Luo, Kisii and Kipsigis home areas. The KTGA was not optimistic about the scope for increasing skill levels and output in this environment, endorsing a rule of thumb of one and a half workers per acre, and it was sceptical about the opportunities for mechanisation of tea plucking and other field work.[39]

Managing Change

The late 1950s were marked by a burgeoning of welfare paternalism as the tea companies recovered from wartime stringencies and made renewed efforts to stabilise their workforce in response to the Carpenter Report. With the Kenya Tea Company and African Highlands in the lead, there was large investment in housing as bachelor accommodation was converted into family housing, to encourage workers to bring their wives and children to Kericho for the duration of the contracts, and to renew them regularly. Soon, every estate village had its primary school, medical dispensary, welfare hall and sports ground. Weekly rations of maize flour and sugar were provided, with school milk for children. These facilities were underpinned by a welfare department, central hospital and even a mobile cinema. Each estate was under the control of a British manager and one or more assistants, themselves subject to a strict conventional regime at work and beyond. Skills for managers and workers alike were learned on the job from more experienced hands.

These material improvements in living standards were then impacted by two external developments.[40] New ideas on management were being promoted back in England around the notion of joint consultation between management and employees on workplace issues. The Kenya Tea Company decided to introduce workers' consultative committees to every estate, in the face of much local scepticism, but the initiative paved the way for more meaningful developments later on. At the national level, the introduction of trade unions to the colony had been actively promoted by the post-war government. In Kenya, trade unions were the only form of organisation permitted for Africans to join during the Mau Mau emergency, and this had been ably exploited, especially under the leadership of Tom Mboya. Then, in 1960, there was the shock announcement of imminent political independence – which was achieved in 1963. Agricultural workers were slow to form trade unions, but the Tea Plantation Workers Union was registered in 1961 and recognised by the tea companies. The Kenya Tea Company now had an industrial relations function as well as a welfare one.

It was soon apparent that the labour force preferred a more arm's length relationship and that cash in hand trumped paternalism. Benefits were consolidated into pay. Meantime, government assumed responsibility for the staffing of schools (with the companies still providing housing) and parents paid the standard school fees. With each workplace having its union representative, there was a somewhat bumpy settling-down process of establishing – and operating – dispute procedures. The tea companies agreed to collect union dues through the payroll, but they did not agree to a closed shop.

Following Kenya's political independence in 1963 the tea companies realised that they would have to give up the model of management by expatriate Britons in favour of employing Kenyans, notwithstanding their ownership structure. To begin with the companies promoted field headmen, or *niarparas*, to managerial positions, but their lack of formal education proved an obstacle as the companies evolved more detailed monitoring of the production process, and they were phased out in favour of the recruitment of Kenyans with higher education qualifications, as well as to the advancement of existing staff who had shown potential. In turn, there was now an emphasis on skills training, both within the companies and through external courses. By the turn of the century James Finlay could claim that all supervisors had a college diploma or better.[41] Foreign ownership by major international investors (Unilever, Swire, Linton Park) has resulted in the transfer of knowhow and global standards to the country. This has been most evident in the scientific improvements to tea cultivation and manufacture, with consequent improvements in yields and productivity. It has also been evidenced by a huge improvement in safety standards in field and factory.[42] The AIDS epidemic in East Africa has presented a particular challenge, with an infection rate in Kericho at one time of 14 percent, and has resulted in substantial treatment and educational outlays.

Evidence for the steady weakening of ties to 'home' in the rural areas is provided by the annual labour census returns compiled by the Kenya Tea Growers Association. In the 20 years to 2000 the Kericho estates increased their planted tea area by 30 percent and their total workforce by 25 percent to 37,000. During this period the number of married resident workers increased by half to 14,600 and accounted for a growing proportion of the workforce, rising from 36 percent in 1980 to 43 percent in 2000.[43] In step with this, the numbers of dependants on the estates rose sharply to nearly 52,000. A further indication of the stabilisation of the workforce was the increased number of workers employed for five years or more, which rose from 4,700 to 13,400 in 2000, when they accounted for 39 percent of the workforce as compared with 17 percent in 1980.[44]

A more serious long-term concern for the tea industry relates to its labour-intensive cost structure, in an environment of weak global demand and prices. The political nature of industry wage awards in Kenya has been insensitive to their long-term implications for employment. All the major companies are evaluating and implementing procedures to reduce labour in the growing and manufacture of tea through pest and fertiliser controls, mechanical harvesting in place of hand plucking and the introduction of streamlined factory procedures. The search for improved quality is also driven by the necessity to improve productivity.

NOTES TO CHAPTER 2

[1] The account is drawn from Thomas, op. cit.

[2] MP/Labour/1, KTGA memorandum to the Labour Commissioner, 8 February 1938.

[3] MP/Labour/1, Conference minute, May 1927.

[4] MP/Labour/1, Memorandum, 15 November 1926.

[5] Ibid., Labour officer, Kisumu to headquarters, 11 November 1936.

[6] Almost every settler household in Kenya in those days (and often up until the 1950s) had its Cinderella in the form of the ubiquitous kitchen boy, or *toto*.

[7] MP/Labour/1, Senior Commissioner to Canon Rogers/CMS Mbale, 4 January 1927. The Kenya Missionary Council resolution of 29 March 1927.

[8] *Report of the Employment of Juveniles Committee*, Nairobi, 1938.

[9] MP/BB/2, Population Returns.

[10] MP/Labour/1, KTGA minutes, 13 August 1938.

[11] Ibid., Letter to *Manchester Guardian*, 10 December 1938.

[12] Ibid., to Colonial Office, 8 December 1938.

[13] MP/KTGA/1/1. The tea labour force in Kericho in 1941 amounted to 13,343 of which 5,713 or 43 percent were juveniles. Census return 31 December 1941.

[14] MP/Labour/1/1, KTGA minutes, 26 January 1943. Afterwards, the labour commissioner wrote to the chief secretary, "I am certain that if the Registration of Juveniles is introduced during

the war, (the KTGA) will be forced to apply for conscript labour to replace the majority of juveniles at present employed, and this will certainly cause the Government some embarrassment owing to the shortage of native manpower at the present time. I am strongly of the opinion that this Ordinance should not be brought into force during the war." Letter, 30 January 1943.

[15] Ibid., KTGA minutes, 25 April 1949. When forwarding the proposal it was argued, "Children will be forced to remain in the Reserve for schooling and to help cultivate their parents' *shambas*. Employers and Government will not be faced with the many difficulties involved in obtaining parents' permission in order to employ juveniles, and control generally should be much easier." Letter to Labour Commissioner, 29 April 1949.

[16] Ibid., Letter to KTGA, 3 May 1949. He went on, "I hold very strongly and have always held that employment of juveniles under decent conditions such as are obtained on tea plantations, is to the benefit of the entire community. It teaches the children routine and discipline; it enables them, or should do so, to contribute to the family budget and, on top of this, we could impose some degree of regularity that is the first stage of disciplining our future labour force – a matter of really paramount importance to the future of the Colony."

[17] Ibid., Report by the Deputy Labour Commissioner on juveniles in Nyanza, May 1959.

[18] Ibid., Record of a discussion held at Kericho at a meeting held on 3 May 1950.

[19] An important lacuna was filled at this meeting. For the first time, the companies had to submit lists of juveniles who had been employed outside the recruitment departments, which meant that they could now be traced by their families. The repatriation guarantee was not extended to this category, but the Labour Department undertook to do this. The meeting also agreed to raise the minimum employment age to 13 years.

[20] MP/Labour/1/1, Part of a response by Brooke Bond to a questionnaire from the European elected members, 29 May 1953.

[21] *Report of the Committee Appointed to Enquire into the Question of the Introduction of Conscription of African Labour for Essential Purposes,* 1942.

[22] Thomas, p.44 and MP/Labour/1, KTGA minutes, 16 January 1945.

[23] Thomas, pp.46–68.

[24] KTGA minutes, 16 January 1945, and circular letter to members dated 28 November 1944.

[25] Ibid., 31 March 1947.

[26] Ibid., 25 November 1947.

[27] MP/Labour/1/1, Memorandum by Thomas, 2 November 1950.

[28] Ibid.

[29] MP/Labour/1, Wilkinson's '*Memorandum on African Budgets'* was tabled at a subsequent meeting on 18 December 1950.

[30] MP/Labour/1, KTGA memorandum for Board of Agriculture meeting on 23 November 1950.

[31] MP/Labour/1, KTGA minutes, 18 December 1950.

[32] MP/Labour/1, Memorandum to KTGA, 12 December 1950. The rest of the paragraph is a prescient vision of future developments in Kericho. "Under native methods of agriculture the statutory two acres allowed to a Resident Labourer is not enough; under improved and controlled methods it should be a considerable asset to his budget; whether enough to de-tribalise him and settle him on the Grower's land will remain to be seen. Controlled agriculture should be followed by cooperative marketing, establishment of better livestock etc. If this plan were followed in consultation with Government the tea areas would really be relieving the pressure on the Reserves, contributing to the welfare of the Colony as a whole and providing themselves with at all events a minimum resident labour force."

[33] *East Africa Royal Commission 1953–55 Report,* HMSO, Cmd 9475. In its discussion, the report did not recognise the plantations sector as presenting any distinctive issues as regards resident labour and the relevant analysis relates to urban workers. Also, *Report of the Committee on African Wages,* Government Printer, Nairobi, 1954.

[34] MP/Labour/1, Deputy chief secretary to Kenya Tea Board, 1 May 1954.

[35] MP/Labour/1, Special meeting of KTGA, 3 September 1952.

[36] MP/Labour/1, Record of conference in Kericho, 7 October 1952.

[37] MP/Labour/1, Government Wages Committee, Questionnaire and Replies. KTGA, 19 June 1956.

[38] The post-war sequence of wage increases by the KTGA up to this point was as follows:

	Basic wage signing-on rate per ticket Shs	Increase %
1944	12	
1947	15	25
1949 end	18	20
1950 end	24	33
1954 end	28	16.7
1954 amended	32 (cash for blanket)	33

[39] MP/Labour/1, Responses to Questionnaire for the Committee on Rural Wages, 19 September 1955. Employment of the three main tribes was Luo, 10,062; Kisii, 8,214; and Kipsigis, 4,204.

[40] MP/Labour/5, I am indebted here to a memoir written in retirement by Brooke Bond's industrial relations manager, Brian Cranwell, entitled *Consultation and Industrial Relations.*

[41] Somewhat paradoxically, Europeans who took out Kenyan citizenship remained eligible for posts designated for localisation. The tea companies have been able to benefit from this dispensation with a number of key management posts. Another paradox was that agricultural properties owned by prominent Kenyans were still allowed to employ expatriate managers in many cases. President Moi's Kaisugu estate in Kericho was so managed for many years.

[42] For example, the perennial danger in Kericho of injuries and deaths from lightning strikes has been largely eliminated by dispensing with iron-roofed plucking shelters and by close weather monitoring such that work is abandoned in the face of impending thunder storms. Another example has been the mechanisation of log splitting for the factory boilers, in place of men with axes.

[43] MP/KTGA/1/1. The census returns only analyse the status of so-called Ungraded Employees, leaving out Graded Employees, who numbered 2,786 in 2000. These people would almost certainly be in married accommodation and of long service. Adding them into the calculation raises the proportion to 41 percent in 1980 and 47 percent in 2000.

[44] Again, adding in the Graded Employees increased the proportion from 24 percent in 1980 to 44 percent in 2000.

3

SOMETHING IN TEA

The circumstances of plantation tea cultivation – the extensive area of monoculture and its remoteness from other centres of population – led to the emergence of distinctive social communities of planter managers and a resident labour force attracted to work on the estates. In Kericho this community flourished for not much more than 40 years between the 1920s and 1960s, before adapting to the realities of independent Kenya. With the replacement of expatriate British managers by Kenyan nationals possessing professional qualifications unknown to the original planter community it vanished. This section goes in search of that departed world, with the help of memoirs from a group of Brooke Bond staff recorded in the 1990s; a memoir of the 1920s recorded by the Pickford brothers in 1971 and supplemented by Joyce Pickford in 2004; interviews in 2003–4 with former staff of James Finlay and Brooke Bond; and, finally, some personal recollections of my own of a Kericho childhood.

Given that the knowledge and practice of tea growing were transferred from India and Ceylon, it was inevitable that elements of the lifestyle of the Indian tea estates should also have been adopted, albeit adapted to the less grand conditions of colonial East Africa. In Kericho, Finlay's first superintendent, William Lee, was transferred from south India in 1925, as were his assistants Fitz Villiers-Stuart and Hugh Thomas. Villiers-Stuart succeeded Lee as Superintendent and he is remembered for having a cruel streak with animals and men: for riding his horse through pruned tea which slashed their legs, and once personally administering a flogging to a factory clerk.[1] Brooke Bond started with the Ceylon experience of Bell and Murray Clarke, and other men with Indian experience were later transferred to Kericho: Jock Dagleish, Bill McConnell, Captain Brindley and Stanley Mills. However, Brooke Bond appointed as their first superintendent Captain Tom Derby, one of the Beadoc farmers who had been appointed to wind up the scheme. He had personal resources of his own and was able to build up a notable racing stable. He had an Irish temper and became a dominant personality, remaining in Kericho in retirement until his death and, more than anyone, he and his wife were responsible for setting Kericho's social style.

After the war there was an initial wave of ex-servicemen, such as Tom Grumbley, who had been in the RAF and stationed in Kenya. On being recruited by Tommy Rutter, who told him "All right, when you are demobbed we will offer you a job in Kenya", the difficulty was how to get out to East Africa. Grumbley remembers being told "Right, well, when you can get out, get out." Then I heard the Uganda Company were wanting to get these tiny planes out to Uganda, so I said I will fly one out for them... The thought of flying to Africa was certainly an adventure, and fun too."[2] He and a friend flew two Austers (which had a range of 200 miles) the 5,000 miles out to East Africa. Later recruits were more likely to be public schoolboys after their national service, all of whom had to be interviewed first by a director before travelling out to Kenya. Francis Monck Mason was the son of a Kent farmer who applied through the Public Schools Employment Bureau, as did John Popham, who later went on to manage Kaisugu estate after it had been bought by President Moi. In Scotland, James Finlay were able to draw on a tradition of overseas employment as well as family connections. Thus, Lindsay Stone-Wigg was the son of an indigo planter in India, Peter Robertson's father was Colombo manager, John Low's father was in the Indian agricultural service and George Corse, David Strachan and Peter Ragg all had tea-planter fathers in the company. The standard interview questions – "Do you hunt? Do you shoot? Do you fish?" – attested to the significance attached to a healthy outdoor life rather than to qualifications. On the whole, Brooke Bond recruited from a wider social circle, reflecting its more down-to-earth origins, and a saying did the rounds: "Brooke Bond staff are planters trying to be gentlemen and Finlay's staff are gentlemen trying to be planters."

As each estate was developed there would be a manager's house and garden, with more modest structures for assistant managers and engineers. They were initially built of timber, then in grey stone with a red- or green-painted roof of corrugated iron. They were often well sighted on a ridge in a preserved clump of indigenous forest trees, but this was not always the case in the early days, as Grumbley recalled:

> "We were given this little shack on the top of the hill, called Kapsakum House... It was literally a corrugated iron cubicle really, divided into four. We were not allowed electricity because Sheila Dagleish said their voltage might be affected in their house below... It was extraordinary, we lived in that tiny house and we had two houseboys, we had a cook, we had a kitchen toto, we had a shamba boy and we had an ayah. That was six people to look after us three in the tiniest house – it was ridiculous."

The Kericho climate and its altitude, between 6,000 and 7,000 feet, were favourable for gardening, so that not only could keen gardeners satisfy their nostalgia for English roses, carnations and sweet peas, but they could also grow the wonderful bulb-and-shrub flora of South Africa. The working day started early, with the labour muster on the estates at 6.45am, which was preceded in the manager's house by the knock on the bedroom door and the arrival of early morning tea. In the early days there was a horse allowance for getting round the estates. Work finished by 4pm, which gave the prospect of some exercise before sunset at seven, except that for much of the year there was a high probability of a thunderstorm in the afternoon. For the wives, mid-morning tea parties were a prime source of social life and district gossip, with much appreciation of cake-making and flower-arranging skills. Children played outside in the garden, under the casual supervision of *ayas*, if they were young.

Following Indian custom, the club was the real centre of social life. Brooke Bond built a first club for its staff in 1927, with tennis courts and a nine-hole golf course to follow. A new and grander building replaced it in 1931, which is still there today. Under Derby's influence, polo, a race course and an annual gymkhana were introduced in 1933. Rugby and cricket were provided for. Wednesday and Saturday were club nights, with attendance being virtually compulsory for the former. The bar was for men only, where Wallace, the rotund Kikuyu barman, officiated genially for many years, whatever the conversation. Bar fights were not unknown, as were tales of weekend-long drinking bouts. In the main club room the company was mixed and there was a radiogram for informal dancing, as well as for special occasions. Whist drives, amateur shows, Scottish dancing and the very occasional political meeting were all part of the scene. Until a church was built in the 1950s, a visiting padre officiated at the occasional church service.[3] On major occasions the parked cars would have beds made up in them for children, who would play around in the dark until tired, under the benevolent gaze of the club's nightwatchman, Arap Wassa.

The Kericho estates became somewhat run-down during the war as a number of managers had been called up for service and those left behind approached retirement. In 1941 there were only 39 managers left on the Kericho estates, which made for a close knit community. There was no new estate development on account of the International Tea Restriction Scheme. However, the social scene was a gay one since many families provided holiday breaks for servicemen from the Middle East theatre. With the ending of the war, the restrictions on development lapsed. Both Brooke Bond and James Finlay moved into development mode and a new investor from India, George Williamson, appeared on the scene. The companies recruited new staff and there followed a 20-year period which marked the flowering of tea society in Kericho. At its peak there would have been around 150 expatriate tea posts in Kericho,[4] which had also become an important district centre for the colonial administration and a township with a number of Indian-owned stores, branches of the British banks and a native hospital. A church was built for the community and Brooke Bond financed the Tea Hotel in order to relieve the pressure on managers to accommodate visitors. It was built without a bar, reflecting the teetotal austerity of the chairman, but bowed to Kenyan realities by conceding the building of a bar in the grounds that was linked to the hotel by a covered way. It became a rival centre of social life in the district and displaced the ramshackle Dandelion Hotel of the 1920s. Writers on the Kenya scene visited Kericho, such as Elspeth Huxley and Negley Farson, as did members of parliament from Westminster and the East African Royal Commission in 1953. Writing in 1948, Elspeth Huxley commented, "Everything about a tea estate seems clean and tidy – even genteel. The factories are spick and span, the roads well kept, the managers well dressed, their wives *comme il faut* and given to afternoon calling, the gardens colourful and well kempt – Kericho is a great place for gardening – and the tea itself, a pretty but rather smug little bush (it is kept little by pruning) covers the steep red hillsides without a single miss or break, a gleaming carpet of glossy leaves."[5]

The season between October and March became known as directors' weather – winter in Britain but the best time of year in Kericho – when head office' directors would undertake tours of inspection. Such visits could last for several weeks as estates were inspected in detail, with culvert covers whitewashed in anticipation. Whatever the events of the day, a round of dinner parties and receptions had to be laid on by evening. The most successful hostesses and engaging company were not necessarily aligned with the most capable managers seen during the day, so that social tensions were inevitable. The Brooke Bond visitations after the war became notorious, when the leadership of the firm was dominated by the thrusting personality of its deputy chairman, Tommy Rutter – or TDR as he was referred to – who was himself the son of the Indian sales director who had visited Kenya in 1914. Despite his notorious temper, Derby was a weak superintendent and TDR stepped into the power vacuum. He dismissed ruthlessly so that, ahead of his visits, speculation would mount as to who was likely to be on the hit list. On a famous occasion a marked man was mistaken by Rutter for one of his blue-eyed boys and thus survived his factory inspection, only to be laid low the following day when Rutter met his real protégé. The company doctor, Douglas Dixon, was summarily dismissed for having the temerity to seek a pay rise; he was replaced by the government doctor from the local hospital, and Douglas Dixon in turn took his place there.[6]

Kericho was far from having the louche reputation of Kenya's Happy Valley set. Bourgeois respectability was the touchstone, although the 1920s were more colourful. One of the original farmers in Kericho, Ian Orchardson, took a Kipsigis wife and had a family, which excluded him from Kericho society. He stayed on with a small tea property until he eventually sold to Brooke Bond and played an important role in the construction of the first Brooke Bond factory; he also did a hydroelectric survey for James Finlay. More sensational was the arrival in Kericho from India in the 1920s of a divorced lady, Mrs Rowe, and her teenage daughter. She set up home with Derby, causing a local scandal, until a degree of respectability was established when her daughter married John Pickford and she could stay with them until herself marrying Derby.[7] Lolly Derby had a strong personality and quickly became the *burra mem sahib* of Kericho, dominating its social life and laying down its sartorial standards for events and dinner parties. Being invited to a morning tea party was like a royal summons, and for children there was a large garden to explore and to hunt for feathers from her peacocks. Disrupters of domestic harmony, or errant husbands, were dismissed "for the good of the company", and both James Finlay and Brooke Bond sacrificed able staff on the altar of respectability. Drink was a more tolerated aberration, and most people had their treasured anecdotes of "swinging on the chandeliers" or – more likely – driving into the ditch on the way home from a party. With so much respectability, sex with African women was virtually unknown.

The tea community tended to be disengaged from the often hectic politics of settler Kenya, and in any event the tea companies would have disapproved of overt political engagement by their managers. In the 1950s, when attempts were made to forge multiracial politics in Kenya and the Capricorn Africa Society was in vogue, Kericho was represented in Legislative Council by Agnes Shaw, the wife of a popular Sotik farmer who had a small tea holding.[8] She was a supporter of Michael Blundell, the leader of the United Kenya Party, which sought to give expression to these multiracial views and to secure an ongoing political role for the white community. Her meetings at the club appeared to enjoy tolerant support from members. In 1952 a state of emergency was declared in Kenya and for some five years British-financed military operations grappled with the Mau Mau rebellion. The emergency mainly affected the Kikuyu people in Central Province adjacent to Nairobi; however, Kikuyu workers elsewhere in the country were mostly displaced and made to return to their ancestral homes. As few Kikuyu had been employed in the Kericho area, the emergency did not impinge on economic and social life as much as in other parts of the country. During the emergency, a number of younger European staff were called up to serve in the Kenya Regiment and the community suffered a trauma when the son of a popular couple was killed on operations.

The plantation lifestyle was destined for drastic change with Kenya's independence in 1964. It quickly became apparent that work permits for expatriates would become an issue and that the tea companies would have to rethink the make-up of their management. This cannot have been a surprise, in the light of their experiences in India and Sri Lanka; nevertheless, little preparation had been made. James Finlay and Brooke Bond pursued contrasting strategies. Brooke Bond set out to negotiate a phased run-down of their expatriates over a ten-year period,[9] whereas James Finlay declined to make such a plan and contested the reductions step by step. The result was much the same for both companies, but there was a sharper drop in morale at Brooke Bond. Many managers had their careers terminated prematurely and did not find it easy to apply their experience to other walks of life. The companies did not provide retirement pensions for their staff; instead, they accumulated a lump-sum provident fund which was meant to provide an annuity on retirement. In the circumstances of the 1970s the fund and its income turned out to be meagre. One consequence was that a number of planters felt that they could not afford to retire to Britain and so remained in Kenya – either on the coast or near Nairobi – where they might hope to continue in a job for a while. Both companies suffered arbitrary cancellations of the work permits of individuals who did not fit into the new Kenya. The outcome was that the tea companies now only have two or three expatriate posts each and there is a new social order of Kenyan managers on the estates, mostly with professional qualifications. There is a preponderance of managers with local links in the area, with the result that nearly everyone also has a family farm with a tea holding.

NOTES TO CHAPTER 3

[1] MP/1M/3/6, Peter Robertson interview 14 February 2004.
[2] MP/1M/1/3, Grumbley memoir.
[3] I was christened, age five, at home on Kimugu estate by the bishop of Mombasa during one of his parochial tours.
[4] Grumbley, op. cit.
[5] *Last Chance in Africa*, Negley Farson, Gollancz, 1949. *The Sorcerer's Apprentice*, Elspeth Huxley, Chatto & Windus, 1948, p321.
[6] MP/1M/5, Joyce Pickford interview, February 2004.
[7] Ibid.
[8] The Capricorn Africa Society was set up by Colonel David Sterling, founder of the SAS regiment, when he went to live in Southern Rhodesia after the war. It was a movement that aimed to change the politics of Central and East Africa by enabling all those living in these countries to play a part in their governance, regardless of racial origins. It influenced the political climate in the years leading up to independence but was overwhelmed by the force of African nationalism. In Kenya, the United Kenya Party gave expression to its aims. I was an active member of its Northern Rhodesian counterpart, the Constitution Party. C.f. *Capricorn*, Richard Hughes, Radcliffe Press 2003.
[9] Grumbley, op. cit.

4

IMPERIAL SOLIDARITY

By the 1930s the tea companies in Kericho were beginning to master their labour recruitment problems as they pressed ahead with planting up their new estates. However, development was brought to a halt in 1933 when the East African colonies were compelled to join an international scheme which aimed to reverse the damaging fall in tea prices that had occurred since 1929. The mechanism the scheme used was restriction on the quantity of tea exported to world markets and a ban on new planting. World exports of black tea, 98 percent of which came from India, Ceylon and Indonesia, had run significantly ahead of consumption and prices had fallen to unremunerative levels for many producers in those countries. The application of the Restriction Scheme to East Africa led to tensions within the KTGA, and between Kenya planters and those in Uganda, Tanganyika and Nyasaland. It was also unwelcome to the Governors of these colonies, who resisted the Colonial Office line, as well as falling out amongst themselves at perceived unequal treatment. The Restriction Scheme was a blatant attempt to serve the interests of long-established tea-producing countries at the expense of more efficient producers and new entrants to the market.

Two earlier restriction schemes had failed, mainly due to Dutch objections on behalf of their Indonesian tea interests. However, the

International Tea Agreement (ITA) that came into force in February 1933 was signed by representatives of the industry in India, Ceylon and Indonesia and was backed by legislation by the respective governments so that it was binding upon all producers in those countries, unlike the earlier schemes.[1]

The main feature of the scheme was that control was to be achieved through regulation of exports rather than production. Partly this was because tea was then seen as primarily an export crop, and exports are far easier to control than production providing there is government backing. There was also a deliberate intention to provide an incentive to develop local consumption of tea in tea-producing countries, which had been very much neglected up until then. It was a crucial feature of the Restriction Scheme that, although governments were not signatories to the Agreement, its success depended upon their active participation in order to licence acreages, give legal sanction to export control and ban the export of tea seed. The Restriction Scheme was a rescue operation for the established industry; so far as possible, new entrants to it were to be prevented.[2]

Once agreement had been reached between the three main participants, attention was turned to securing the adherence of the lesser tea-growing countries to the scheme. Their cooperation was not essential to the success of a tea restriction scheme, but it was felt to be desirable on moral grounds and to prevent these fringe areas from exploiting their position once prices began to improve by rapidly extending their tea acreages. It was thought to be particularly important that the tea-growing British colonies should join the scheme, since the well-being of a major Empire industry was at stake. Accordingly, in February 1933, the Indian and Ceylon Tea Associations approached the Colonial Secretary, who then addressed the governors of Kenya, Uganda, Tanganyika and Nyasaland, explaining the objectives of the scheme and seeking their views on its application.[3]

The Secretary of State's despatch did not evoke an enthusiastic response from the local governments. The Tanganyika government was preoccupied with the problems of struggling planters, who had been encouraged to turn to tea growing, and the governor concluded: "it is clearly the duty of this Government to do its utmost to encourage the successful establishment of a tea industry in the Territory."[4] The Governor of Uganda thought that no action was required: "The Agricultural Department and private individuals are concerned with exploring the possibilities of the industry rather than with production."[5] In Kenya the government had immediately asked the KTGA for its views. Since this body was dominated by two companies with large tea interests in India and Ceylon, it could be expected to take a favourable attitude to restriction; indeed, the chairman of the International Tea Committee (ITC) responsible for administering the scheme was also the chairman of James Finlay. While agreeing with the main objects of the scheme, important modifications were suggested for its application to East Africa: there should be no restriction on exports, but Kenya growers would agree to cease all development providing that those growers who had just started development were allowed to complete economic units. The following formula was proposed:

"(a) that planters who have 100 acres or more under tea, and who have no means of disposing of their leaf to the larger factories, should be permitted to extend to a minimum economic area of 500 acres including a fully equipped factory; (b) that small estates and individual growers who sell their leaf to a central factory, may enlarge their areas to a maximum of 100 acres each."[6]

A necessary condition to these proposals was that they had to be accepted by all the East African governments.

The KTGA's reasoning was that, in the circumstances, East Africa regulation based on standard exports was impracticable, since the immaturity of most of the tea estates left no proper basis for calculation and, in any case, exports had hardly started. In any event, the position in East Africa was the reverse of that pertaining in India and Ceylon. There, the problem was overproduction from a large established acreage, so that exports were the crucial factor;

in contrast, exports from East Africa were negligible and the policy aim was to arrest development through the control of new planting. A further consideration that weighed with local producers was the potential for domestic consumption. Although an optimistic front was maintained about the prospects for such trade, the fact was that home consumption was not growing despite competition and low prices, and producers did not relish the prospect of being compelled to sell a fixed proportion of their crop to the local market. The justification for allowing the completion of economic units was that a number of small growers would be caught out if all development was to be stopped immediately. It was felt to be necessary to request this concession in order to win their backing for the Restriction Scheme.

It was agreed that the forthcoming East Africa Governors' Conference in October 1933 would be the occasion to reconcile the differing viewpoints, despite the fact that Nyasaland, which was not a member, was absent from the conference and there was no link between the respective tea interests at that time. However, the ITC persisted in dealing with East Africa on a four-territories basis throughout the period of tea restriction – a procedure that had exasperating consequences for East Africa.

The Governors were in a difficult position. As they wrote to the Colonial Office: "East African Governments feel bound to develop such East African industries as are possible within their territories, but recognise that it is undesirable for increased production in East Africa to militate against the policy of tea export regulation adopted by India, Ceylon and Dutch East Indies."[7] These countries, through the ITC, which they had set up in London, took a hard line with the Colonial Office: "We are already seeing increased quantities of tea coming into the world's markets from these Dependencies, and we do not see why producers in East Africa should ride on our backs to take advantage of a situation which is created by a scheme such as is now in existence; in fact we are definitely of the opinion that only controlled production can save us from falling into a worse position than we have already been in, and we consider we have a claim on British connections to assist us in this matter."[8] They had already enforced a ban on all exports of tea seed, which was a powerful sanction because only very small quantities of seed were available in East Africa and its quality was suspect – a material consideration for a crop with a life expectancy of 50 years or more.

The ITC did agree to accept the framework proposed by the KTGA of freedom from export controls in return for a cessation of new planting, other than an allowance to establish viable economic units for those who had already started planting. However, it demanded that this be accommodated within a total new planting acreage of 6,000 by 1938, rather than the aggregate bids of 11,000 acres (including Nyasaland) that had been submitted by the governors.[9] The issue was remitted to the next Governors' Conference meeting in April 1934 to sort out the territorial allocations, and the delegates in turn turned to the industry representatives, committing themselves to accepting and enforcing their advice. The Kenya Government in particular came to regret this step.

The KTGA delegation won a considerable victory by persuading the other parties to accept their allocation proposals and to keep within the target of 6,000 acres.[10] However, this took no account of Nyasaland.

These decisions were forwarded to the ITC through the Colonial Office, which added a figure of 2,000 acres for Nyasaland. By this time – June 1934 – the overriding consideration of the ITC was to get the East African governments to sign up as quickly as possible. The Colonial Secretary was therefore informed that, in view of the urgency of reaching finality, the claims would be accepted, even though they were regarded as far too high, and provided there was no further appeal. "The Committee press this point because it would appear to be possible that Kenya, Uganda and Tanganyika may demand further increases when it has been realised that Nyasaland has been allowed 2,000 acres."[11] On these terms, the East African colonies joined the ITA. Ordinances were passed in the legislative councils and tea seed imports from India were allowed once more. The outcome of these protracted negotiations may be summarised as follows:[12]

	Kenya	Uganda	Tanganyika	Nyasaland
1932 acreage	12,000	700	2,000	12,000
Expansion bid	4,000	2,900	5,000	3,000
Approved by ITC	1,000	2,000	3,000	2,000

The Deputy Director of Agriculture attended a meeting in December 1934 to apply the terms of the award. It was agreed that estates providing green leaf to a factory should not exceed 100 acres, which disposed of two applications from Limuru (whose growers were not represented on the KTGA at this time). He was then persuaded to agree that no planter with less than 100 acres should be eligible for an allotment either as this would favour small producers too much, although this contradicted the formula as originally proposed. The result was that only factory estates were eligible for allotment and allocations of the 1,000 acres were made to four small estates in Kericho, to the outgrower estate in Limuru owned by Brooke Bond and to an estate in Nandi.[13] It subsequently became apparent that the 1,000-acre claim formulated by the KTGA had not actually contained a bid for small green-leaf suppliers to complete 100 acres, nor had it made full allowance for the small factory estates to build up to 500 acres, and the Department of Agriculture subsequently calculated that a further 2,308 acres could have been included. When it is recalled that the small outgrowers supplied leaf to Brooke Bond's factory in Limuru, it illustrates the influence exerted by India on the major companies to conform to the restriction policy.

The most constructive aspect of the Restriction Scheme was a worldwide campaign to expand tea consumption, which the partic- ipating countries financed by a cess (a local tax) on production. A body called the International Tea Marketing Expansion Board was set up to administer this promotion drive. The ITC was naturally anxious that the East African territories should participate in this aspect of the scheme as well, but the idea did not find favour. It was felt to be an unjustified burden on a very young industry, many members of which (particularly in Uganda) were not even interested in the export trade. The KTGA was strongly opposed to the idea when it was first discussed in November 1933, but opinion was more divided a year later and it gave rise to a more constructive thought, "the idea of a cess on tea produced or exported should be opposed but that everything that was possible should be done to increase the local sales and consumption of East Africa tea."[14] An officer of the International Tea Marketing Expansion Board visited East Africa in 1936. He achieved no results in Uganda or Tanganyika, but the KTGA agreed to levy a cess on production for propaganda purposes, providing the proceeds were spent locally. The Kenya Government agreed to pass the necessary legislation, which came into effect in January 1938.

The Tea Cess Board levied a cess of 80 cents per 100 lbs of made tea and a promotion campaign was initiated in the native reserves, starting with Nyanza.[15] The campaign had barely got under way in 1939 when war was declared and fieldwork became impossible; it was closed down in 1940, save for a little ongoing advertising. The Tea Cess Board was re-convened in 1947, by which time funds amounting to £36,000 had accumulated. But it was not a propitious moment, since tea was in short supply and producers were anxious to export as much of their output as possible at the more favourable export prices. Plans to set up an East Africa-wide bureau came to nothing since the Kenyan growers refused to support funds to expand local consumption while tea prices were still subject to official control. In 1949 the International Tea Marketing Expansion Board withdrew altogether from East Africa.

Estimates submitted to the ITC of growth in local tea consumption proved overoptimistic. Although only one producer (Brooke Bond) had an effective sales organisation, the KTGA forecast a doubling of local consumption over four years, and this was raised to a trebling at the 1934 conference, notwithstanding that tea was a new beverage to the native population. Consumption actually increased by some 40 percent.[16]

Scheme Renewal

Halfway through the restriction period, the ITC decided to renew the scheme for a further five years, from 1938–43. Having secured the support of the main producing countries, attention turned to East Africa and negotiations were opened in August 1936 directly with the tea associations in Kenya and Nyasaland, with the Colonial Office being kept informed of developments. This caused some offence to the governments of Uganda and Tanganyika who felt, with some justification, that they were being hustled into a fait accompli. More than two years of negotiations followed before a settlement was reached, in which the ITC repeatedly had to modify its offer in order to accommodate the vigorously expressed views of governments in East Africa. The Colonial Office was little more than a postman. The governments were responding to local political pressures, but in Kenya the situation was complicated by the realisation that the KTGA was too much the voice of the big tea companies, whereas there was a wider constituency to be accommodated of failed coffee farmers wishing to switch to tea growing.

The ITC wanted East Africa to renew as full members on the same basis as the main producers as regards nil new planting, but with the export quota being calculated by reference to estimated future production, rather than past performance, to take account of the immaturity of the tea plantations. When the question of renewal was put to a meeting of the KTGA in September 1936 the chairman (from African Highlands) proposed acceptance of the terms, since "African tea growers had benefited very materially from the higher tea prices realised as a direct result of the present scheme, under which the regulating countries were bearing a very heavy burden."[17] However, there was strong opposition and an extraordinary general meeting was called to vote on a rival resolution asserting that the Association "is not prepared to undertake future participation in the new scheme in respect of regulation of exports." It was argued that the industry had already made sufficient sacrifice in holding back on development, "both as regards its own interests and the interests of the Colony as a whole" but, more importantly, the local market could not be realistically expected to absorb the quantity of tea that would become available in the coming years from the existing acreage if the export quota system was applied to Kenya.[18] The opposition resolution was carried by a two-to-one majority.[19]

The reason for such vigorous opposition from a body that normally saw eye to eye with the ITC lay in the state of the local market, where prospects were being assessed much more soberly than in 1934. Consumption was rising very slowly and competition from tea growers in Uganda and Tanganyika was increasing. It seemed as if sales of Kenya tea might remain static at around 2 million lbs per annum, while production was expected to rise to 12.5 million lbs by 1943. It was feared (incorrectly) that the new Restriction Scheme would only allow 60 percent of production to be exported, with the result that large quantities of tea would have to be held back. It was asserted at the meeting that Kenya might have 3.7 million lbs of unexportable tea in 1938 and 5 million lbs in 1943. It was significant that the Kenya Tea Company was in the opposition camp.

Six months later, emollient influences had been at work and the KTGA was able to pass a resolution stating that it "would be prepared to give sympathetic consideration to any proposal which would, in view of the infancy of the African Tea industry, extend towards it a more lenient treatment than that to be applied to the older established producing countries."[20] In response, the ITC suggested that exports be restricted to 80 percent of potential production, but this was still considered too harsh, since the standard export quota on the main producing countries was 87.5 percent. The difficulty was that the quotas for the main producers were fixed in relation to a past base year, whereas the East African quotas were fixed in relation to forecasts of potential future production.[21] A further compromise was proposed: the future production estimates should be cut by 5 percent and then an export quota of 87.5 percent applied. This equated to an export quota of 83 percent of forecast production and it was accepted by the KTGA in September 1937, on the stipulation that the ruling was accepted also by the other East African territories.[22] The effect of the new proposal was to set Kenya's proposed export quota for 1938 at 7.67 million lbs on estimated production of 9.25 million lbs, leaving 1.57 million lbs to be absorbed by the local market. However, in December 1937, the ITC raised the export quota for all producing countries to 92.5 percent, effective from 1938. For Kenya, where the production estimate had been revised to 10.8 million lbs, the new quota figure was 9.5 million lbs, leaving only 1.3 million lbs to be sold in the local market. Nyasaland accepted the ITC's proposal but only at this point were the three East African governments asked to endorse them and the carefully constructed compromise collapsed.[23]

The Uganda Tea Association, newly formed in 1936, resolved to request 1,400 acres for new planting during the renewal period to enable existing growers to complete economic units. This was supported by the government on the grounds that tea was regarded as a product for the domestic market rather than for export, and the Governor of Uganda, Sir Philip Mitchell, spent part of his overseas leave successfully lobbying the Colonial Office and the ITC for the additional planting, minuting that he "had a hideous battle in London to get this concession."[24] The Tanganyika Government pressed similarly for an allowance for new planting to meet domestic demand.

The Kenya Government had its own agenda, comprising both a grievance and a demand. The grievance was over the 1934 acreage allotment, where the failure of the KTGA delegates to apply for the full allotment to which Kenya was entitled – at the expense of the small growers – needed rectifying in the eyes of the Department of Agriculture. It calculated that, under the formula, the Kenyan application should have been for 2,308 acres, instead of the 1,000 acres actually applied for. The government was also resentful of the special treatment accorded to Nyasaland and was jealous of the way that Uganda and Tanganyika had been allowed to manipulate the economic acreage formula to secure a greater share than was strictly intended. There was perhaps an element of sour grapes since the Kenya government had let itself be guided by the KTGA at the time, and had not taken steps to find out the composition of the figures put up at the Nairobi conference in 1934 until sometime later. Nor had the Colonial Office vetted the figures, even though the ITC had protested at the size of the claims of Uganda and Tanganyika, and had only agreed to them "provided that the Secretary of State is satisfied that these can be justified."[25]

The new Kenya proposal, as summarised by the Acting Colonial Secretary, was a controversial one: "There is the separate question of the development of the Colony as a whole in the interests of its inhabitants, as distinct from the development of an industry by Companies whose major activities lie outside this Colony and on whose interests a small extension of area of planting in Kenya will have little or no effect... Tea is now known to be an economic crop in certain areas where it had not proved itself (in 1934), whereas other crops have proved a failure in places where there is good reason to expect that tea would afford certain planters a living which they can gain from no other branch of agriculture. Several applications have been made recently by persons who desire to plant tea in suitable areas which have failed to respond to development under other crops, e.g. in the Nandi and Kaimosi areas, and the Government desires in the interests of the Colony to support these applications."[26] The government had received petitions from several planters' associations to be allowed to switch from unprofitable coffee planting to tea planting, and felt that their pleas were irresistible. The Kaimosi Soldier Settlers had applied for 1,000 acres, the North Sotik Farmers Association for 1,000 and the Nandi Planters Association for 3,000, with the last body concluding their application on a political note: "The Colonists and Government have recently affirmed in no uncertain terms their belief in white settlement, and have pledged their support to farmers in legitimate efforts to place their farms on a profitable basis. Government would not deliberately embark on a policy of 'freezing out' the individual farmer or planter."[27] After examining all the claims, the government resolved to apply for 2,200 acres.[28] It had no hope of support from the KTGA in this matter, since the Association had only voiced the viewpoint of the larger planters, who had nothing to gain from new entrants, and it was anyway too much under the influence of parent companies with interests in India and Ceylon.

A situation had emerged in which only Nyasaland and the KTGA had accepted the ITC's proposals and some concession was essential if East Africa was to be brought into the renewal agreement. The ITC first came up with an arbitrary acreage for new planting of half what had been allowed in the first five years, which amounted to 3,400 acres and was intended to allow small estates to come up to an economic size. However, this did not properly address Kenya's historic grievance and a second proposal of 4,700 acres was made, and loftily endorsed:

> "The Secretary of State, as was his predecessor in 1934, is impressed with (the argument against new entrants) and trusts that, particularly in view of the generous offer now made by the International Tea Committee, your government will be prepared to waive its demand for an additional acreage for new entrants into the tea industry. He regards the maintenance of the scheme as of great importance to one of the major Empire industries and would not feel able to support any proposal which might jeopardise its continuance."[29]

Kenya was still unsatisfied, as the offer only restored the 1934 oversight and did not provide for new planting on former coffee estates, on which issue the farming community and the European elected members were by now fully alerted. Following a debate in Legislative Council, the Acting Colonial Secretary was able to write: "The Government is in the fullest sympathy and would be faced with considerable difficulty in securing the passage of any Bill on the conditions proposed by the International Tea Committee."[30]

A further concession was called for and the ITC duly agreed to offer another 2,300 acres to East Africa, but without specifying that it was for Kenya or for failed coffee planters, as the same argument could be made for Tanganyika – and was. The allocation problem was remitted to the three directors of agriculture in April 1938, but their proposal failed to secure endorsement by the Governors' Conference and the hot potato was returned to London, pursued by despatches. The figures were rearranged to favour Kenya and the East African governors accepted the award. However, Nyasaland objected and made direct representations to the ITC, who granted it 1,000 acres and then a further 500 acres to East Africa as a consolation prize. "The concessions outlined represent the Secretary of State's final word."[31]

After some two and a half years of negotiations, and from a starting point that there should be no new planting or new entrants, the ITC had eventually been persuaded to allow for 8,500 acres of additional tea planting, as follows:

	Kenya	Uganda	Tanganyika	Nyasaland	Total
Dec 1937	1,300	1,000	1,400	1,000	4,700
Aug 1938	1,000	275	525	500	2,300
Dec 1938	200	175	125	1,000	1,500
Total	2,500	1,450	2,050	2,500	8,500

These allocations were made with certain conditions, which were summarised as follows:

a. Export regulation based on estimated production, but standard exports to be defined more strictly by making a deduction of 5 percent from production for the first two years, 7.5 percent for the next two years and 8.75 percent for the fifth year.

b. A propaganda cess on production to be increased in stages to Shs2 per 100 lbs of made tea.

c. The new planting to be spread over five years, with not more than 40 percent planted in any two years.

d. "The admission of new entrants to be confined to cases where interests or individuals already established in the territories have themselves tried reasonable acreages, or hold land on which trials have been made of other crops which, owing to soil or climatic conditions, have failed to become economic undertakings."

e. Allocations must be in economic units for factory organisations.[32]

In the final settlement, Kenya had been allocated 1,300 acres to make good the shortfall under the first scheme to complete economic units and 1,200 for new planting by former coffee growers, where the Department of Agriculture was sympathetic to farmers who had been affected by the combination of coffee berry disease and low prices and were seeking an alternative livelihood. The issue was acute in Kaimosi, Nandi and Sotik – all areas where tea growing appeared to be a viable alternative. Little consideration seems to have been given to the capital investment implications of a four-year waiting period before harvesting could start, or to the aspect of factory costs. Strict application of the criteria for 500-acre economic units resulted in only 746 acres being allocated to three estates in Kericho (Kaisugu, Mau Forest, Kapkorech), to Mabroukie in Limuru and Mokong in Nandi; and 210 acres were allocated to eight growers to bring their holdings to 100 acres: seven at Limuru and one in Kericho. This left 1,544 acres for allocation to new planters, where the government had a delicate problem in the light of bids for some 4,500 acres. In the event, 625 acres were allotted to eight farmers in Nandi, 330 to 11 farmers in Sotik, 300 to ten farmers in Kaimosi and 60 to two farmers in Kericho. This left a balance of some 200 acres for further small allocations to individual farmers. The Director of Agriculture summed up the exercise in a speech to the Nairobi Chamber of Commerce: "These areas in Nandi, Sotik, Kericho and Kaimosi are those in which coffee estates have been severely attacked by coffee berry and allied diseases. The allotments have all been made to settlers and not to companies. With one exception the allottees are all new entrants to the tea industry."[33]

Thus, after much effort, the East African governments were brought into the renewal of the Restriction Scheme for the period to 1943. Elsewhere in Africa, the government of Southern Rhodesia had declined to join the scheme, and the Belgian and Portuguese governments had likewise refused to bring their colonies into the scheme. In reviewing the experience of the first restriction period between 1933 and 1938, the ITC was candid in saying that it had not met early expectations, especially that it had only been possible to increase export quotas from 85 to 87.5 percent in the last year. Nevertheless, prices had doubled since reaching a nadir in 1931–32. Perhaps its most significant reflection for the established tea industry was the observation "It has greatly helped to re-establish the industry, and it has brought back prosperity to many concerns, who in 1932 were facing ruin. Others it has saved from possible heavy losses."[34] The ITC was sure that the Restriction Scheme should be renewed again and the outbreak of war did not alter this view, notwithstanding the fact that quotas had now been raised to 95 percent.

NOTES TO CHAPTER 4

[1] For a study of the ITA see *Tea Under International Regulation*, VD Wickizer, The Food Research Institute, Stanford, CA, 1944.

[2] The main provisions of the Agreement were:
  - That tea exports should be regulated in order to restore the balance between supply and demand.
  - That the governments of the producing countries should undertake to prohibit exports in excess of agreed quotas.
  - That the basis for regulation should be the maximum exports reached by each participating country in either of the years 1929, 1930 and 1931.
  - That the initial quota should be 85 percent of the standard figure, this figure to be re-examined annually by a special standing committee, after reviewing the world statistical position.
  - It would be a condition of participation that no new planting should take place (guaranteed by governments), and that tea seed exports to non-participating countries would be prohibited.
  - That the agreement should be for a period of five years.

[3] MP/ITA/1/1, Secretary of State to Governor of Kenya, 3 March 1933.

[4] MP/ITA/1/1, Governor of Tanganyika to Secretary of State, 5 May 1933.

[5] MP/ITA/1/1, Governor of Uganda to Secretary of State, 3 May 1933.

[6] MP/ITA/1/1, KTGA minutes, 2 June 1933.

[7] MP/ITA/1/1, Governors' Conference, cable to Colonial Office, 10 October 1933.

[8] MP/ITA/1/1, ITC memorandum presented to Secretary of State, 21 December 1933.

[9] MP/ITA/1/1, ITC to Secretary of State, 23 March 1934. East Africa had claimed an aggregate of 8,000 acres, of which Kenya accounted for 4,000 acres and Nyasaland for 3,000 acres.

[10] The application of the formula to Kenya resulted in cutting its bid from 4,000 acres to 1,000.

[11] MP/ITA/1/1, ITC to Secretary of State, 6 June 1934.

[12] 1932 acreage figures (rounded) from International Tea Committee, MP/Library/13; expansion bids as set out in Governors' Conference cable to Colonial Office, 10 October 1933; final allocations Governors' Conference note 21 April 1934.

[13] In Kericho: Kaisugu was owned by Brayne, Kapkorech by Lord Egerton, Mau Forest by Grant and Chelimo by Caddick. The Nandi allocation to Mokong owned by Mayers was subsequently revoked on his death and reallocated to the others.

[14] MP/ITA/1/1, Meeting of Kenyan, Ugandan and Tanganyikan producers in Nairobi, 29 April 1934.

[15] MP/KTGA/2/5b. This paragraph is based on the minutes of the Tea Cess Board. The cess was set to raise some £4,300. The advertising slogan in Swahili announced 'Drink Tea It Makes You Strong'.

[16] MP/ITA/1/1, KTGA Memorandum 1933, See also MP/Library/26 my B.Litt. thesis, p139

[17] MP/KIA/1/1, KTGA minutes, 15 September 1936

[18] MP/ITA/1/1, KTGA minutes of Extraordinary General Meeting, 29 September 1936.

[19] African Highlands was supported only by Mau Forest and Kapkorech, while ranged against it were the Kenya Tea Company, the Buret Tea Company, Jamji, Orchardson and all the Limuru farmers.

[20] MP/ITA/1/1, KTGA minutes, 2 March 1937.

[21] MP/ITA/1/1, KTGA minutes, 29 June 1937. The latest estimates of potential production tabled at the meeting forecast 12.1 million lbs in 1938 and 14.4 million lbs in 1943.

[22] MP/ITA/1/1, KTGA minutes, 3 September 1937.

[23] A fuller account of the renewal negotiations is contained in my thesis, MP/Library/26, p141ff.

[24] MP/ITA/1/1, Minute of 25 January 1938.

[25] MP/ITA/1/1, ITC to Colonial Office, 6 June 1934.

[26] MP/ITA/1/1, Acting Colonial Secretary to Governors' Conference, 15 November 1937.

[27] MP/ITA/1/1, Memorandum by Director of Agriculture containing a comprehensive review of the claims from associations and individual planters in Nandi, Kaimosi, Sotik, Kericho and Limuru, June 1938.

[28] Of the 8,290 acres applied for, 6,715 acres were in respect of unsuccessful coffee – an indication of the over-reliance on this crop in these districts.

[29] MP/ITA/1/1, Secretary of Governors' Conference forwarding the Secretary of State's despatch, 28 December 1937.

[30] MP/ITA/1/1, Acting Colonial Secretary to Secretary Governors' Conference, 28 January 1938.

[31] MP/ITA/1/1, Colonial Office to Secretary to Governors' Conference, 8 December 1938.

[32] MP/ITA/1/1, 8 December 1938.

MP/ITA/1/1, Statement on the allocations by the Director of Agriculture to Nairobi Chamber of Commerce 14 July 1939.

[34] MP/ITA/1/2, ITC Annual Report for 1937–38.

5

EAST AFRICA BREAKS FREE

It might have been expected that the outbreak of war would have led to the lapse of the Tea Restriction Scheme. The loss of most European markets only accounted for some 2 percent of total imports for consumption, whereas the elimination of tea exports from the Dutch East Indies from 1942 removed 17 percent of world exports for the duration. However, this was not the view of the ITC, which recommended to governments that the agreement should be kept in force during the war and for two years afterwards.[1] The Colonial Office was of the same cast of mind: "In general it is expected that after the war there will be a wide extension of principles of international control of primary products," officials wrote to the Governor of Kenya in 1942, "and industries such as the tea industry which can show that they have a scheme in being with a long record of success will be much more likely to be left to operate independently... It is the opinion of His Majesty's Government that all existing schemes should be continued for a sufficient period after the termination of hostilities to enable the question of their further continuance to be reconsidered at leisure. They have recently agreed to the renewal of the tin scheme for 6 years from 1st January last and for the continuance of the sugar scheme from 1st September next, and the rubber scheme from 1st

January next for a suitable period. They could not therefore consistently assume a different attitude in regard to tea."[2]

The news was not well received. It had become clear by now that East Africa had a comparative economic advantage over India and Ceylon by virtue of its much higher yields per acre (1,000 lb or more versus around 400 lb) and that tea estates were very profitable at current prices. As the Kenya Director of Agriculture wrote: "Ceylon and Indian tea interests in the UK were able to secure the imposition of this agreement with a view to maintaining in production estates which under normal circumstances were uneconomic."[3] Moreover, it was correctly noted that the loss of production from the Dutch East Indies had transformed the market situation, while there was resentment that Belgium and Portugal had refused to join the Restriction Scheme in respect of their African colonies.

The KTGA was conflicted on account of the presence of the interests of Brooke Bond and James Finlay, but sentiment in the colony was encapsulated in a motion tabled by the European elected members in Legislative Council: "That this Council recommends to the Imperial Government that all restrictions on the planting of tea in the East African territories should be abolished in view of the destruction of tea estates in the Far East."[4] The Director of Agriculture stated that the government was pleased to accept the motion. While conceding the general merit of international commodity controls, Kenya did not wish to maintain the present one as a model for future development and a proposal was forwarded to the Colonial Office that 1,000 acres of new planting be authorised every year during the war, either for existing or new planters, and that the agreement should lapse six months after its end.

The ITC thought it wise not to engage in a prolonged argument this time and it offered East Africa new planting up to a total of 20 percent of the already permitted total as at March 1943. This worked out as 5,665 acres, distributed as follows:

Kenya	Uganda	Tanganyika
3,232	943	1,490

Moreover, it also agreed that any unplanted balance of the previous allocation could be carried over, in response to pleas that several small growers had been called up for military service and had been prevented from fulfilling their quotas.[5] These proposals were accepted by the East African governments and the ITA was extended for the duration of the war and for one year thereafter.

For Kenya, the ITA settlement offered three years of expansion at the rate proposed by the government, i.e. up to 1946, but serious differences arose as to how the additional acreage should be allocated. The majority view in the KTGA was that the 20 percent expansion should be distributed pro rata to existing acreage, but there was vocal opposition from small growers.[6] However, the government was determined to encourage new entrants to the industry. The Kenya Tea Company briefed solicitors to support its case for the pro rata approach, which would have allowed it a further 822 acres of tea on top of its existing area of 4,110 acres.[7] When it learned that the government was proposing to favour new applicants it joined with African Highlands in writing to the colonial secretary in protest, arguing that this would lead to land speculation rather than development and that capital costs of £150 an acre would be beyond the capacity of individuals to finance. Oliver Stanley, Secretary of State for the Colonies between 1942 and 1945, declined to become involved in the matter. In Nairobi, the government set up a committee to determine the basis of allocation, with guidelines that pointed to the desirability of opening up new areas and encouraging small growers.[8] Its decision was to allocate 39 percent to new growers (1,270 acres), 30 percent to small growers to complete economic units (955 acres) and 31 percent between established interests (1,007 acres).[9] Under this dispensation the Kenya Tea Company would have been allotted one-third of their expectation, or 274 acres.

In East Africa the 1943 renewal of the Restriction Scheme was for the duration of the war plus one year. At an officials' meeting in 1946 the termination date was deemed to be 31 March 1947 and it was decided to hold an inter-territorial conference in January 1947 to consider what should happen next.[10] Two days before the meeting, the KTGA passed a resolution making several points: control over new planting was no longer necessary and should be opposed; the principle of export regulation was accepted; the ITC should continue to exist as a statistical body, pending such time as export control might again be needed.[11] At the inter-territorial meeting it was agreed to lapse planting restrictions (which was followed up in March by removal of export quotas), and that the Tea Ordinances should be amended accordingly. This was done by Kenya and Uganda, but Tanganyika allowed its ordinance to lapse. In July the ITC took the view that, by these actions, East Africa had withdrawn from the Restriction Scheme and, as a result, exports of tea seed to East Africa were banned by India and Ceylon.[12] In November the ITC decided to extend the ITA to 1950, without enforcing export quotas for the time being, but with the removal of all restrictions on new planting.[13]

Notwithstanding the ban on seed exports to East Africa, the ITC was anxious to retain these countries as members. Exploratory discussions took place during the autumn and an official visit to East Africa was contemplated. It appeared that the ITC had an exaggerated view of the scope for new tea development in the region, as a threat to the established industry in Asia.[14] However, a new factor had impinged on the scene, namely the international negotiations for the establishment of the General Agreement on Tariffs and Trade and an international trade organisation. This led to a complete turnaround in the attitude of the Colonial Office, as it realised that the Tea Restriction Scheme was contrary to the letter and spirit of the new world trade order that was emerging. More specifically,

> "The UK would welcome the continuation of the ITC as a Bureau of Information and Statistics but would not agree to the clauses which –
> 1. Prevented the export of tea seed by countries party to the Agreement to countries which were not party to it;
> 2. Prohibited the planting of tea except under licence (this did not include the licensing system operated by territorial authorities for agricultural reasons only);

3. Restricted the quality of tea a producing country could export."[15]

However, this did not prevent the newly independent states of India, Pakistan and Ceylon from entering into a new five-year agreement in 1950, but colonial East Africa was no longer a party to it.

Appraisal of the Restriction Scheme

In the nine years leading up to the launch of the Restriction Scheme in 1933 there was an enormous increase in tea planting in the traditional areas of India, Ceylon and the Dutch East Indies, where an additional 347,000 acres of tea were planted. The resulting increases in production and exports from these countries peaked in 1932, with exports to world markets 18 percent higher than in 1924. It is scarcely surprising that auction prices dropped to a low point in 1932, at half the level they had reached in 1927, especially when it is borne in mind that tea consumption is not price-sensitive and more reflects the social conventions under which it is regularly drunk. Both acreage expansion and export growth were arrested by the ITA and prices immediately improved markedly, albeit not to the previous high point. In this perspective, the ITA achieved its primary aim of rebalancing supply and demand through its action on the three main producing countries, and this was reinforced by the loss of Dutch East Indies' production during the war.

The ITA also correctly discerned that stimulating the consumption of tea in world markets was part of the solution to oversupply, and it established an International Tea Market Expansion Board. This well-meaning body had little real impact on the market, and its operations were suspended during the war. Much more influential was the pressure induced by the Restriction Scheme to stimulate tea consumption within producing countries. Here, the presence of Brooke Bond in both India and East Africa, as an experienced tea marketing company, was fortuitous. India increased its consumption of tea nearly sevenfold between 1938 and

1950 to 97,000 tons, at which point it was equivalent to 53 percent of its export volume. In Kenya, on a much smaller scale, consumption rose fourfold between 1935 and 1950 to 3,000 tons, which was then equivalent to 45 percent of its export tonnage. In both countries the tea-drinking habit had become widespread.

Given the statistical insignificance of East Africa in the world tea picture in the 1930s, it is surprising that so much effort was devoted to drawing the colonies into the ITA. From a later perspective there seems to have been a lack of proportion, as well as ill-concealed hard feelings. The chairman of the ITC, Sir Robert Graham, wrote a review of the working of the scheme in 1942 and commented, "In passing it may be mentioned that an element which British representatives have always found very embarrassing in their relations with their Dutch colleagues is the fact that British tea interests outside India and Ceylon always betrayed a desire to take advantage of the Scheme and in this some of them were at times supported by their local Governments... The difficulties experienced in respect to the East African territories and Malaya were all the more disappointing because, as British interests in tea in these countries predominated, a more sympathetic attitude to the Scheme had been expected."[16]

It is arguable that world economic conditions in the 1930s would have naturally checked the tea acreage expansion that had been taking place in Asia, and that weaker producers would have been forced out of business. Be that as it may, the advent of the ITA had a decisive effect in rebalancing global supply and demand and this was achieved more quickly than would otherwise have been the case. It was of course unfortunate that all this happened just when the potential of East Africa had become apparent.

The tea estate companies that were established prior to the Restriction Scheme had their development severely constrained during its lifetime. However, once war had been declared it must be doubtful whether the outcome could have been very different, since the estates had then to contend with severe manpower problems. The companies had to release managers for war service and were reduced to a skeleton staff; meanwhile, they experienced acute labour recruitment difficulties due to the competing demands of the military. In these circumstances it seems clear that the planting constraints of the Restriction Scheme were not a material constraint on expansion during the war and in its immediate aftermath. Nevertheless, this early phase of development clearly demonstrated Kenya's comparative advantage as a tea producer, which flowed from the longer harvesting season and much higher yields per acre. While South Asia averaged 200–400 lbs per acre, Kenya was demonstrating 1,000 lbs per acre and better. This underlined the strong sentiments from the KTGA and the government not to be constrained from new development once the war was over.

The Agriculture Department underestimated the suitability of tea growing as a crop for failed coffee farmers: the cost of cultivation over a four-year waiting period before maturity and the capital costs of a factory were beyond the means of financially straightened settlers. The model of Brooke Bond's Limuru factory and its farmer outgrowers had not yet registered as a solution to the problem, so that the planting allocations that were argued for so strenuously largely went to waste.

At the imperial level, the India Office clearly trumped the Colonial Office when the interests of India and Ceylon clashed with those of the African colonies. Notwithstanding vigorous lobbying from colonial governors, the colonies eventually had to fall into line with the strategy of the ITC. In any event, the industry position was compromised by the fact that James Finlay and Brooke Bond were part of the India/Ceylon tea establishment and had to behave accordingly. In Kenya, the political clout of the settlers was significant in strengthening the views of the colonial administration, but this also led to tension within the KTGA, where four of the smaller estates were locally owned.

The ITA was an industry-driven initiative that required governments to pass enabling legislation to enforce its rules on new planting and on exports. The corporate links between James Finlay and Brooke Bond with African Highlands and the Kenya Tea Company provided powerful leverage for the prevailing orthodoxy. Nevertheless, the ITC still had to work through the Colonial Office and the East African governors and their local legislative councils in order to get its way. In reflecting on the negotiations to bring East Africa into the ITA and on the subsequent renewal negotiations one is led to conclude that – given the importance attached to bringing East Africa into the Restriction Scheme – a failure to master sufficient detail and understanding of the workings of colonial administration made for contentious outcomes. To begin with, the ITC appears to have had an exaggerated impression of the threat represented by the East African colonies, in terms of the potential for its new planting to undermine the Restriction Scheme unless it was arrested, or at least controlled. It appears to have made no systematic effort to assess this potential in terms of land suitable for tea planting, let alone in terms of there being parties with the means to undertake significant development.

In Kenya, James Finlay and Brooke Bond had secured virtually all the potentially available land in Kericho and, of course, were firmly aligned with the party line of the ITC. The four independent estates companies in Kericho — Buret, Egerton, Kaisugu and Mau Forest— did not have much additional land and were financially constrained, as were the tea farmers in Limuru. The potential of Sotik, Kaimosi and Nandi had yet to be established and the farms were mostly in the hands of settlers with little capital, who had already suffered from unsuccessful coffee growing ventures. Such information was obtainable through the Finlay/Brooke Bond connection, but was also officially sent to London through the Governors' Conference in 1933 and 1934.

Against this background, the ITC endeavoured to enforce the total ban on new planting that had been applied to the major producing countries, and then became embroiled in a series of compromises in an effort to keep new planting to as small a figure as possible, in which the notion of equitable treatment between the territories was lost. All the colonial governments were reluctantly prepared to accept the principle of the Restriction Scheme, but in return they expected accommodation of their relatively modest development aspirations. In practice, what happened was that the ITC surrendered to pressure as it failed to implement an even-handed allocation. The official wish list for the whole of East Africa was for just over 15,000 acres of new planting by 1939, but it did not require much scepticism to conclude that, with the exception of the Kericho companies, this was unlikely to be achievable in the circumstances of the time.

The problem for Kenya in 1933 was that the government had allowed the KTGA to formulate the policy response: no restriction on tea exports and new planting only to allow existing growers to complete estate units, or to reach 100 acres if selling green leaf to a factory. The big tea companies had agreed to cease new planting and no consideration was given to tea growing elsewhere in the colony. This approach was acceptable to the ITC, but the Uganda and Tanganyika governments had meanwhile formulated their bids based on aspirations for a new agricultural sector. The ITC did not engage with their proposal, but instead imposed an arbitrary limit of half the acreage requested, leaving it to the colonial governments to sort out the territorial allocation. At the same time, a side deal was cut with Nyasaland. The result, as we have seen, is that development in Kenya was effectively arrested in the period leading up to the war, while Uganda and Tanganyika were able to use their allocation to make a start with promoting new development.

Unsurprisingly, the first renewal of the Restriction Scheme from 1938–42 involved two and a half years of contentious haggling as far as new planting was concerned, following the ITC's opening offer to allow each colony half of its first-round allocation, without regard to what had happened on the ground. The Kenya Government was determined to rectify its failure to spot the way that the KTGA had interpreted its mandate, as well as to do something for failed coffee planters. This opened the door to special pleading by the other governments. The ITC was unwilling to get involved in the detail and dealt in arbitrary concessions, which earned it the hostility of the Kenya Government, as well as that of governments elsewhere. It was only with great difficulty that the Kenya Government was persuaded to renew the Restriction Scheme in 1943 for the duration of the war, and it withdrew from it at the first available opportunity afterwards.

The principal effect of the Restriction Scheme on the Kenya tea industry was one of delayed development by the two main Kericho companies. In the early 1930s they were both engaged in major expansion, having the land, finance and capability. This came to a halt in 1933. The tea acreage of African Highlands remained at 4,593 acres from 1933–47 and that of the Kenya Tea Company in Kericho at 3,614 acres. When they were eventually able to commence new planting, labour shortages and the government of India's ban on seed exports were both serious initial constraints. Meantime the capital costs of development had escalated from under £100 an acre and were rising towards £500. The efforts of the Department of Agriculture to allow room for a white-settler tea sector in areas where coffee growing was unsuitable were unsuccessful. The farmers lacked sufficient capital to finance the preparatory phase, let alone factory construction. The example of Limuru – a factory estate with small outgrowers selling green leaf to it – was only taken up some 30 years later in the very different circumstances of African smallholder tea growing.

NOTES TO CHAPTER 5

[1] MP/ITA/1/2, ITC to Secretary of State, 1 January 1942.

[2] MP/ITA/1/2, Colonial Office to Governor of Kenya, 13 May 1942.

[3] MP/ITA/1/2, Chief Secretary to Secretary Governors' Conference, 7 March 1942, enclosing the memorandum.

[4] MP/ITA/1/2, Motion tabled on 21 April 1942.

[5] MP/ITA/1/2, Cable from Colonial Office quoted in KTGA minutes of 13 July 1943.

[6] MP/ITA/1/2, KTGA minutes, 13 July 1943.

[7] MP/ITA/1/2, Letter dated 14 December 1943. It argued in part, "In our opinion, a very strong point in our favour is that this company owns 19,000 acres of proved tea land, and had it not been for the fact of Restriction on further planting being imposed in 1933, we would actually have proceeded with development and, at a conservative figure, should now have had approximately 9,000 acres under tea."

[8] MP/ITA/1/2, Statement on allocation for the Kenya Information Office fortnightly bulletin 26 July 1944 for publication in August.

[9] MP/ITA/1/2, KTGA minutes, 6 March 1945.

[10] There was a legal puzzle in that the termination date was to be one year after His Majesty the King had officially stated the war had come to an end. However, no such declaration had been made; hence the decision that the legislation should be allowed to lapse on 31 March 1947. Ibid., Meeting note of 6 October 1946.

[11] MP/Library/26, KTGA minutes, 7 January 1947, quoted in thesis p.171.

[12] As it happened, there was a serious outbreak of blister blight disease that year in India and

Ceylon and the East African governments decided early in 1948 to prohibit all seed imports in order to protect the industry. Thereafter, East Africa had to rely on its own tea seed resources.

[13] MP/lTA/1/2, ITC minutes, 12 November 1947.

[14] MP/lTA/1/2. There is an extensive file note of a discussion on 23 October 1947 between Morrison of ITC and the EA tea controller, D.S. McWilliam, in which Morrison notes concerns about the way that acreage allocations had been made; the doubts that local growers had of the benefits of the Restriction Scheme; the view that the ITC had greatly exaggerated the potential for acreage expansion; and, finally, urges a visit to the region. The meeting was subsequently referred to by the commissioner Eastern Africa Dependencies, R. Norton, in a letter to Tanganyika of 12 March 1948 with the comment "the ITC appeared to have gathered many erroneous impressions".

[15] MP/lTA/1/2, Note of meeting with Sir Gerard Clausen, 31 March 1948.

[16] MP/lTA/1/2, *Review of the Tea Regulation Scheme 1933-43*, Sir Robert Graham, London, 1942. In a communication to the shareholders of James Finlay, Sir Robert was even more outspoken, as reported in the *East African Standard*, 31 July 1933. He accused East African interests of trying to sabotage the Scheme with the backing of the local administrations "which might have been credited with greater breadth of vision. East Africa entered the international scheme as latecomers on generous terms, and it was impossible to conceive that growers in that part of the Empire could have any grievance as far as the crop basis is concerned. When the full significance of the present demand to be permitted to plant out new areas of unrestricted acreage was appreciated, involving as it does, the breaking of faith with our partners, the Dutch East Indies growers, now happily temporarily dispossessed of their properties, it was to be hoped we should have heard the last of the ill-conceived and contentious movement."

6

BECOMING A TEA-DRINKING COUNTRY

Open Competition and Formation of the Pool

Brooke Bond first entered the Kenya market with the aim of selling tea produced by their Indian company. Recognising the potential for tea growing in Kenya, they acquired a large acreage in Kericho and set about developing a plantation business targeted at the export market. James Finlay were motivated solely by the potential for exports. The colonial governments were interested in a new line of agricultural income, and tended to think in terms of local market consumption, especially in Uganda and Tanganyika. The East African customs union levied a high duty on imported tea – primarily for revenue purposes, although it had the effect of protecting the local industry, as Brooke Bond quickly noted. The duty of 45 cents a pound (five and a half pence) compared with London auction prices in the 1920s of between ten and 17 pence. As the Kericho tea estates began to mature and local tea increasingly replaced the imported product, customs revenue fell sharply from a peak of £16,755 in 1928 to £423 in 1933.[1] Without any warning to the tea companies, an excise duty on locally made tea of ten cents a pound was introduced with effect from 1932, in which year the customs union raised £5,139 on a quickly rising trend as production expanded. By 1937 its value exceeded £10 million (worth £50,000 of excise duty).

Through the Kenya Tea Company, Brooke Bond was the only planter with a distribution organisation working in the local market and prioritising the expansion of sales, despite the relative attractions of the London auctions, as it explained to the International Tea Committee: "It has been the policy of this Firm to make the local market its primary concern, and it has never deviated from this objective either when an actual loss has been incurred or when, as now, a much larger profit could be obtained by shipping all teas to London for public sale."[2] It was unfortunate that the fall in tea prices, which reached crisis levels between 1930 and 1933, coincided with the arrival of new Kenya-produced tea. Suddenly the local market was more attractive to the growers and a period of intense competition ensued.[3] All the estates appointed agents to sell their teas on a commission basis, including Brooke Bond. The sharpest rivalry in 1930 was between Brooke Bond and the Buret Tea Company, since African Highlands was selling tea ex-factory as well as through its labour-recruiting organisation at a higher price and did not appoint a distributor until the end of the year. In March, Brooke Bond stated they would be willing to purchase the African Highlands' tea crop, an offer which was rejected. Lee's initial reaction was to strike a bargain with Brooke Bond over market-sharing, but he was rebuffed; indeed, it was to take six years to reconcile all the divergent interests at work within the Kenya tea industry.

In 1931 the prevailing price level for tea was 90 cents a pound – African Highlands came into line in June – and this was held despite the introduction of the ten cents excise duty, which was absorbed by the producers. In October, Buret cut its price from 90 to 85 cents and Brooke Bond followed suit. By the end of November, Brooke Bond had reduced its price further to 80 cents, but this level was not held. The local advantage over London prices had been eroded and the head office directors at James Finlay were concerned: "We intend to have our full share – proportionate to output – of the local market for tea," they wrote to Kericho on 30

June. "If other producers are not prepared to accept that position, then local producers will lose the advantage that they have of being within the customs Union and the local price will be reduced to the same price that would be obtainable by selling in London. This is just what occurred in the case of the sugar industry in Natal until the producers combined to sell through one organisation and to send overseas the excess of local production over local consumption. If the market is to be divided out, it would appear that we shall at least have to agree on a fixed selling price, a policy which is not, I think, always satisfactory, and one that in the case of tea is a much more difficult matter owing to different qualities, than in the case of sugar. It would almost seem that we might have to contemplate the possibility of all producers having in the future to sell through one agency."[4]

An exploratory meeting between the managers of African Highlands and the Kenya Tea Company in August made no headway as the chairman of Brooke Bond declined to consider an arrangement over local prices. African Highlands then reduced its price to 85 cents a pound, a move that was followed by the other producers, who succeeded in winning some 25 percent of the market through their agents, compared with the 50 percent held by Brooke Bond. Competition was increasingly manifested through agent rebates, which moved up from 4 to 7 percent. At this point a strengthening in London auction prices eased the pressure on local competition as exports became more attractive. A phase of price leadership by Brooke Bond followed and, in October 1933, it raised the wholesale price from 85 cents to Shs1 per pound, with the other producers following suit. Merchants read the signs and there was a wave of speculative buying in December. The policy over this period was gradually to raise prices to match the equivalent London price level, which made for a profitable local market. In February 1934, Brooke Bond raised prices to Shs1.15 and again in September to Shs1.25 (this time African Highlands were informed in advance of the move). In 1936, the ex-factory production costs of the Kenya Tea Company were 41 cents per pound (before local marketing and distribution expenses).

The Pool Distribution Agreement

By 1936 a new situation had arisen. London auction prices had improved steadily over the preceding three years, removing the pressure to unload tea production onto the local market. Looking ahead, the prospect was more sombre as a consequence of East Africa joining the ITA. Kenya and its neighbours had been excused from the export quota part of the scheme in the period ending in 1938, but when the International Tea Committee opened discussions on renewal of the scheme in 1936 it expected East Africa to conform to the same terms as the major tea-producing countries. This alarmed members of the KTGA who feared that Kenya's rapidly increasing production would give rise to a requirement to retain more tea for local consumption and that this would shatter the precarious market stability – and profitability – that had been established since 1933.[5]

These fears were compounded by a tendency both to overestimate the projected increase in production and, more importantly, to anticipate a more onerous export quota regime than the ITC intended, let alone implemented.[6] Thus at the KTGA meeting in September 1936 it was asserted that the export quota would be only 60 percent of production, which was forecast as 9,250,000 lbs for 1938. This would have implied holding back 3,700,000 lbs of tea for the local East African market, which was thought to be worth about £2 million (consumption in 1935 had been £1.8 million), but holding back as much as £5 million by 1943. As for the export quota, the ITC's opening proposal in June 1937 was in fact 80 percent of forecast production. The subsequent modifications, already reviewed, resulted in a quota regime of 87.9 percent of forecast production in years 1 and 2, 85.5 percent in years 3 and 4 and 84.4 percent in year 5.

The climate was conducive to reviving a scheme for controlling sales distribution in the East African market through quotas allocated to individual producers and setting up an organisation to administer the cartel. In August 1935 preliminary discussions took place between African Highlands and the Kenya Tea Company, while in Britain there were exchanges later in the year between the chairman of James Finlay, Sir John Muir, and Gerald Brooke, chairman of Brooke Bond. "I am not personally unfavourable to the regulation of prices on the local market," Brooke wrote, "It is that I know the difficulties which arise of making a successful arrangement of this nature… I will therefore send out by this mail a line to our people expressing willingness to come into a suitable arrangement and asking them to furnish us with views which are expressed locally." Muir's response was more robust:

"I think growers in East Africa will be foolish if they cannot come to some arrangement which will enable them to get a reasonable price for their produce sold in Africa, instead of receiving, as is at present the case in Ceylon, what amounts to practically to the cost of production, leaving overheads out of account."[7]

African Highlands and the Kenya Tea Company put two of their executives to work on a scheme and, by November, Fitz Villiers-Stuart and Douglas McWilliam had drawn up a detailed proposal.[8] The essential feature of the proposed scheme was that Brooke Bond would act as the sole distributors of tea in East Africa on behalf of the producers, selling exclusively under its existing brands an equal proportion of the crop of each participating grower. For undertaking this task, Brooke Bond would receive a commission of 7.5 percent of sales. The Brooke Bond sales organisation would act for the Pool, but there would be provision to take on sales agents from other producers if they complemented the Brooke Bond structure. Producers would be required to match Brooke Bond's quality standards. It was proposed that the agreement between producers would be for an initial period of five years. It was also proposed that the agreement should be between Kenya producers in the first instance, but that Uganda and Tanganyika producers would be invited to participate. In summing up the rationale of the scheme, the authors concluded:

"On expiry of the International Tea Agreement in 1938 it is to be anticipated that East Africa will be called upon to reduce its exports on the same basis enforced upon other producing countries. This will imply the compulsory retention for disposal on the local market of a considerable proportion of the output of every producer. Growers with an insufficient or unestablished local trade will be faced with three alternatives:

    a. To sell their surplus tea to an established distributor.

    b. To compete individually for a share of the local trade on a cut price basis.

    c. To restrict plucking.

A scheme on the lines indicated would in all probability provide a market for the entire restriction quota of every grower and assure a price which it would be impossible to expect with uncontrolled competitive marketing."

In Kericho, the overriding concern was over the consequences of a resumed price war between rival producers, as the Kenya Tea Company superintendent emphasised when forwarding the joint proposals to his chairman. "In assessing the advantages from our point of view," he wrote, "it would not be fair to draw a comparison with our present, modestly profitable trade. What we have to consider is the future in the face of unrestricted competition and the danger that, if an Export Quota comes into force, tea will be dumped on the local market at a few cents a lb."[9] But in Glasgow a different concern was now troubling James Finlay. The two chairmen met in early December and Muir then recorded his worries in the following letter to Brooke. "It may be accepted that producers in the East African territories will not agree to Brooke Bond being sole distributors and to the existing branch belonging to B.B. East Africa Ltd being exclusively used, unless they can be effectively safeguarded against BB&Co receiving undue advantage, should any arrangement made between the general body of producers and BB&Co come to an end."[10] This was the preamble to a radically different proposal: that a new producer-owned and -controlled distribution company be established, to which producers would be bound to sell exclusively all tea for the local market. Brooke Bond would be appointed by the company as its selling agents, on commission.

For James Finlay – and presumably for other producers when the scheme was put to them – there was now an issue of confidence.

Producers would have to abandon their hard-won footholds in the local market and place themselves in the hands of their greatest competitor. What if the scheme 'broke down' after a year, or was not renewed? Would its administration be impartial? The perspective was different for Brooke Bond. It held a commanding position in the East African market, now with 58 percent of sales of Kenya tea won for their own producing company. A proportionate share of the market in relation to Kenya tea production would entail the Kenya Tea Company only being entitled to a 35 percent share (based on estimated 1936 production) and having to withhold from the East African market 19 percent of its crop which would otherwise have been sold locally.[11] For Brooke Bond the sacrifice of local market share would only be worthwhile on the condition that it should manage the joint selling organisation, retaining full control of brands and sales policy.

It may be thought that rather than promoting a scheme to share the market, Brooke Bond might have done better to force out competing brands. There were two main considerations that shaped their course of action. First, although it might have been possible to knock out some of the smaller brands, African Highlands were too strong for this tactic to be attractive; it had a significant market share for its *Ndege Chai* brand and the financial strength to withstand a price war. Second, the anticipated renewal of the Restriction Scheme with a 60 percent export quota was expected to force more tea onto the local market and to exacerbate competition. Against this, Brooke Bond had an incentive to foster a market-sharing scheme and to resist a producer-owned sales organisation when it was in a position to lay down the fundamental conditions under which it would operate. It also provided an incentive to try to accommodate other parties and to allay anxieties.

Brooke Bond accepted that the other producers needed a safeguard in the event of termination of the agreement, and proposed that this be achieved by establishing a base date of each producer's market share in 1935, which participants would be entitled to revert to. However, it regarded the notion of forming a new producer-owned company as quite impractical.[12] Both local superintendents were in favour of the scheme as proposed (with the addition of the termination clause), and they both preferred to leave Uganda and Tanganyika out of the scheme to begin with.[13] They felt that the important next step was to get the other Kenya producers on-board, and the proposals were put to the KTGA for the first time in February 1936.[14]

Simultaneously, Brooke wrote to Muir, stressing the modifications that had been made to meet his concerns: the proposed 1935 reference base for market share after termination; the introduction of an advisory committee which would have the responsibility of determining the amount of any cess for tea promotion and would agree on an annual basis the charges to be deducted from the net selling price in arriving at the price payable to producers. He also proposed that producers should agree to adjust their crop each year to the sum of the amount allowed under the export quota under the ITA rules and the Pool quota for the local market.[15] James Finlay's head office in Glasgow was not satisfied at all, and placed on record its fears of market dominance by Brooke Bond. The proposed safeguard of reverting to 1935 market shares was rejected, without proposing an alternative. The scheme was seen now to have a basic flaw: "It appears to us to place altogether too much power into your hands. We are not sure in this connection exactly what authority you would suggest should be given to the Advisory Committee, but even if that Committee were to have the final say in determining prices and policy to be pursued in selling, we feel that the whole selling organisation would be entirely in your hands and that you would be likely to be entrenched in such a strong position as would make it difficult for other producers, particularly small ones, to compete with you."[16]

This train of thought led James Finlay to revert to the notion of a producer-owned sales organisation, managed by Brooke Bond and the letter continued: "We are of opinion that the organisation and the marks under which tea is sold should belong to the producers. We consider that by being assured of a reasonable market for your own tea and of a commission on all tea for a period of years, you would be very well recompensed for giving up any small value that may be in your present goodwill." Finally, it was deemed essential that Uganda and Tanganyika be brought into the scheme from the outset. Another two years elapsed before accommodation of these opposing views was achieved. The issues raised by James Finlay over the potential abuse of power by Brooke Bond and their alternative proposal of a producer-owned marketing organisation re-emerged 40 years later and Brooke Bond eventually lost its marketing role on behalf of the industry. However, on this occasion its proposals prevailed.

There was no immediate response to James Finlay's concerns. Back in Kericho, McWilliam analysed the business tensions within the Kenya market in a lengthy memorandum. He pointed out that Brooke Bond had a significant investment in tea distribution, whereas "no other producer has any appreciable interest to sacrifice in connection with the local trade – some nothing at all. James Finlay's reference to their willingness to give up their mark can carry very little weight, as their trademark on the basis of packet trade is of extremely small value. Bulk trade has no good will and should not be accepted as a point of importance. The main point which James Finlay dwells on reflects their disinclination to allow Brooke Bond to control distribution."[17] On the question of a producer-owned distribution company taking over the Brooke Bond organisation, with ownership based on relative acreages, McWilliam pointed out that the result would be that James Finlay would have the major shareholding interest. He continued:

"Instead of the continuance of distribution by Brooke Bond methods, supplemented by the recommendations from an Advisory Committee, Brooke Bond's wishes and interests could always be vetoed by combined opposition from James Finlay and the smaller producers. Also each and every producer would be in a position to directly interfere with the organisation and running of the new company. In effect Brooke Bond would be called upon to sacrifice their entire individual interests in East Africa as distributors, including their trade marks, as with a new company new marks would be required."

Negotiations resumed in the autumn, with discussion over

strengthened safeguards for members and more operational detail, but unexpectedly James Finlay questioned whether there was any longer a need for a market-sharing scheme:

"The opinion which we have formed is that, with the establishment of a reasonable price level in export markets, owing to the regulation of exports by the three main producing countries, there is little likelihood of proposals for joint selling in East Africa being considered by tea interests there... Provided good profits can be secured by shipping teas to London, producers will argue that there is no occasion to take any very great interest in a scheme which predicates a pro-rata allocation of the teas required in the local market."[18]

This led to the conclusion "We feel that the restriction of exports and the scheme for joint selling are so bound together, we should wish, in bringing the matter before producers in Kenya, to combine the two." The dialogue then lapsed for 12 months, with Brooke Bond perhaps reasoning that it could continue to consolidate its dominance of the domestic market, and in the expectation that renewal of the ITC with export quotas for East Africa was going to happen.

Towards the end of 1937, discussions resumed in London with a revised memorandum from James Finlay.[19] Its introductory paragraphs introduced a new dimension. After noting that the original 1935 paper had significantly underestimated actual production in 1936, as well as the absorptive capacity of the domestic market, and that a driving force of the proposals had been the threat of large forced retention of output, the outlook now was for an export quota of 90 percent, with no pressure on the local market, coupled with satisfactory prices.[20] On the contrary, East African governments might well now be more concerned that enough tea was retained for domestic consumption. The memorandum went on to argue that this factor was now a sufficient justification for a cooperative scheme to share supplies to the domestic market:

"It may be anticipated that the provision of a sufficient

quantity of tea to meet the demands of the local market throughout the period of regulation will be a matter of concern to the Governments of the three territories and that, accordingly, a flexible scheme for the orderly marketing of supplies would be helpful to all producers."

The revised proposals were presented to an extraordinary general meeting of the KTGA in April 1938, where they were met with such critical comments that the meeting was adjourned for a month. When it was resumed, the objections had been smoothed over and it was decided that the agreement should be for an initial period of five years and that the initial quota for the domestic market should be 30 percent of production.[21] A resolution giving formal approval to the Pool scheme (as it came to be called) was passed at the June meeting, with a rider "That Uganda and Tanganyika be approached with a view to obtaining their cooperation, but that the scheme be implemented by Kenya growers in the interim."[22] When the formal document was signed on 8 September 1938 the term of the agreement had been extended to six years. The contentious termination clause now made a distinction between Brooke Bond and the others: the Kenya Tea Company undertook to restrict its sales of tea for a year to the same quantity as it had sold in 1937, whereas, for the other producers, the restriction was to be to sell no more than their Pool quota in the year preceding termination. The effect was that Brooke Bond would have been able to increase its sales above its Pool quota on termination, up to its dominant market share before the Pool was established.

Establishment of the Pool was a victory for Brooke Bond's patient diplomacy, as James Finlay came to terms with its market strength in East Africa in terms of its established network and sales experience. However, it was undoubtedly helpful that the renewal of the ITA in 1938 brought East Africa into the regime of export quotas, even though the initial fears of being forced to restrict production proved unfounded. East Africa was fortunate that, despite increasing production from the maturing estates, the export quota allowance eventually established, plus local demand, happened to be in balance with total output, without any special measures.[23] In Kenya three brands were withdrawn from the market and Simba Chai was dominant. Since Brooke Bond was also selling actively in Tanganyika and Uganda, McWilliam estimated that Pool-grown teas would account for 82 percent of the market at the outset. There was one estate brand in Tanganyika with 5.5 percent of the market and two estate brands in Uganda with 10.5 percent. He was pessimistic about eliminating these brands through competition because he thought the producers would make sacrifices to maintain a foothold in the local trade.[24] Instead, he favoured more negotiation and, in 1938 and 1939, he was the diplomatic intermediary, keeping the other producers informed about the finalisation of the Pool agreement, as well as on Brooke Bond's sales strategy.

Ambangulu estate in northern Tanganyika was keen to join the Pool from the start, and even allowed its supply of packing teas to run down in anticipation of its acceptance as a member. But there was a hitch due to a contract with its sales agents that it was unable to terminate satisfactorily, and so actual membership was delayed until September 1940.[25] In southern Tanganyika the German tea farmers were effectively controlled by the Usagara Company in Germany and negotiations stalled until the outbreak of war, at which point the custodian of enemy property placed the administration of the German estates with Brooke Bond for the duration. Brooke Bond formed a new company, Tanganyika Tea Company, to manage the properties and it joined the Pool in February 1940. In the same year a small tea-packing facility was created at Mufindi. With the agreement of the Advisory Committee, it was arranged that the quotas for the Tanganyika Tea Company should be reciprocal with those of the Kenya Tea Company, so that Mufindi could supply as much of the demand as possible in the southern part of that large country.

Bringing the Uganda producers into the Pool proved to be a much more arduous process and was not finally accomplished until after the war. Brooke Bond had established a sales presence in Uganda before local estate production started. However, two London-funded companies, the Uganda Company and Buchanan's Tea Estates, started their own estates and there was a cooperative venture, the Toro Tea Company, which commenced production in the late 1930s. The initial stance of the Uganda Tea Association in 1937 to the Kenya proposal was that "Cooperative selling under a common packet or trade mark should not be lost sight of, but is somewhat too ambitious a project at present" and that it would be preferable to try to achieve a local price deal between producers, who had been following a strategy of undercutting Brooke Bond in order to gain a share of the market.[26] In March 1938, McWilliam noted that, in Uganda, "there exists a general desire for the crop to go only to the local market. This is understandable when it is remembered that there are no large producers and the existing output is shared by a number of small estates, mostly in private ownership, with neither the wish to export nor the crop and facilities for doing so."[27]

The Uganda Tea Association was favourable to the principle of cooperative marketing, but it was concerned over limiting the proportion of the crop reserved for the local market to 30 percent; Uganda wished to reserve half of its domestic market for its own producers.[28] McWilliam and Thomas went to Kampala and agreement was reached to reserve half of its domestic production for Uganda teas. However, the agreement was not renewed a year later because, in the meantime, the Uganda Company had decided to secure its own share of the market and had set up Uganda Tea Sales (UTS) to compete with the Pool, not just in Uganda, but also in Kenya and Tanganyika, with aggressive rebates to agents.[29]

The Pool had barely a year's normal operation before the outbreak of war, which transformed the marketing outlook. McWilliam's review of its working noted the success in bringing all the Tanganyika growers into the Pool, with the exception of a small Swiss-owned estate. With regard to Uganda, he advised patience rather than a price war in the expectation that, sooner or later, their producers would join the distribution scheme. In response to suggestions that local consumption could be stimulated by price reductions, McWilliam cautioned that experience did not support such an approach, while wartime conditions were so far tending to reduce native purchasing power, on account of higher import prices and reduced employment in coffee and sisal. The best prospect for increased Pool sales lay in enlarging its distribution network beyond East Africa to include Sudan, as a result of disruption of its traditional supply routes.[30]

NOTES TO CHAPTER 6

[1] In 1930 the duty was reduced to 40 cents and then increased to 50 cents in 1931. At the outbreak of war it was raised again to Sh51.

[2] MP/Sales/1/1. Response to questionnaire from the ITC, 3 February 1937.

[3] MP/Library/3. This account owes much to chapter 8 of Thomas's *Brief History*.

[4] MP/AH/1/, Sir John Muir to Lee, 30 June 1931.

[5] MP/ITA/1/1, KTGA Minutes, 15 September 1936.

[6] The KTGA estimates exaggerated the production levels in the early years, but understated those in later years as higher yields became established. For example, production was 850,000 lbs below the estimate in 1940, but 2,117,000 lbs above it in 1942.

[7] MP/Sales/1/1, Exchange of letters 7 and 8 October 1935.

[8] MP/Sales/1/1, *Memorandum of suggestions jointly submitted Messrs AHP and Messrs BBEA Ltd, as a basis for discussion to prepare a scheme acceptable by the Kenya tea growers to stabilize prices and control tea sales on the EA market*, 3 November 1935. Villiers-Stuart devised the main elements of the scheme; he succeeded Lee in 1938. McWilliam was by then managing the sales organisation and, in due course, assumed responsibility for the Pool cartel for the next 20 years.

[9] MP/Sales/1/1, Brindley to Gerald Brooke, 6 November 1935.

[10] MP/Sales/1/1, Sir John Muir to Gerald Brooke, 5 December 1935

[11] These estimates were made towards the end of 1935, using actual sales figures relating to the previous year, and estimated production figures for the following year. Kenya teas accounted for 1.78 million lbs in the regional market in 1934 and Brooke Bond supplied 1.04 million lbs. In 1935 the Kenya Tea Company accounted for 2.3 million lbs of the Kenya crop of 6.6 million lbs and the production estimate for 1936 was similar. If the market had been shared out in proportion to expected production, the company would have had to give up 418,000 lbs of local sales. MP/Sales/1/1, *Memorandum* op. cit. and Brooke Bond head office letter of 12 December 1935.

[12] MP/Sales/1/1, Letter, 12 December 1935.

[13] MP/Sales/1/1, Brindley to Gerald Brooke, 8 January 1936.

[14] MP/Sales/1/1, KTGA circular, 10 February 1936 signed by the two superintendents.

[15] MP/Sales/1/1, Brooke to Muir, 10 February 1936.

[16] MP/Sales/1/1, James Finlay to Brooke Bond, 19 February 1936.

[17] MP/Sales1/1, Notes on G Brooke proposals and correspondence, March 1936.

[18] MP/Sales/1/1, Gatheral to Brooke Bond, 16 October 1936.

[19] MP/Sales/1/1, *Memorandum on Cooperative Selling of Tea in the Customs Union of Kenya, Uganda and Tanganyika*, dated 25 February 1938, which was first sent to Gerald Brooke on 2 November 1937.

[20]

1935 estimates for 1936	1936 actuals
Production 6,430,000	9,265,000
Exports 5,080,000	7,585,000
Domestic 1,350,000	1,750,000

[21] MP/Sales/1/1, KTGA Extraordinary General Meeting, 12 May 1938.

[22] MP/Sales/1/1, KTGA minutes, 28 June 1938.

[23] Customs & Excise figures for exports and local consumption in 1937–39 do not reconcile exactly with agricultural census and KTGA production data on account of transit volumes and stock holding. The Customs & Excise figures are between two and eight percent higher than the production figures.

[24] MP/Sales/1/1, Memorandum by McWilliam written early in 1938.

[25] Ambangulu was given a special Pool quota based on its actual sales to the domestic market, rather than on the Pool production formula, which would have halved its allocation.

[26] MP/Sales/1/4, Uganda Tea Association minute, 10 July 1937.

[27] MP/Sales/1/1, Note of discussion with Sir Theodore Chambers, 1 March 1938.

[28] In 1937, Ugandan tea consumption was 560,000 lbs and its own producers accounted for 275,000 lbs, or 49 percent, so that the request matched the contemporary position. The Uganda Company's share of this output was three-quarters.

[29] The Toro Tea Company was a co-shareholder, and Buchanan joined in 1941.

[30] MP/Sales/1/3, Report dated 4 September 1940.

7

A CUPPA FOR THE TROOPS

While Neville Chamberlain's appeasement diplomacy in 1937–38 was exploring possible accommodation with Hitler, the protracted dialogue also gave Britain more time to press on with preparations for its failure. This applied not only to re-armament, but also to many other consequences of moving the country onto a war footing, which included arrangements for essential food supplies. The tea industry's preoccupation in Kenya with managing distribution in the local market took place against a backcloth of Europe's drift to war, but soon it found itself dragged into the bigger picture. Hitler's Anschluss with Austria was concurrent with the final discussions within the KTGA over setting up the Pool, and the agreement was signed during the dismemberment of Czechoslovakia in September 1938. In London, the Board of Trade began making preparations for war and for the management of essential supplies, which naturally included tea as an essential morale steadier for the troops and on the home front. A Food (Defence Plans) Department was established in the autumn of 1938 and a scheme for the purchase of tea from India, Ceylon and East Africa was circulated. A new role for the Pool was imminent.

East African tea production had to be reconciled between four competing demands: the domestic market; military requirements, large forces being based in East Africa for the Abyssinian Campaign; exports, particularly to neighbouring countries, of which Sudan became the most important; and the Ministry of Food. The Ministry's requirements were known early, but the tea commissioner had no powers to enforce their fulfilment (and exports were an acknowledged priority). The other demands were harder to predict, except that they were growing.

The Pool was designed to share out the protected East African market between its tea producers, and to avoid beggar-my-neighbour competition, at a time when there was expectation that export quotas under the ITA would be insufficient to take up the whole annual crop. By the time that the Pool agreement was signed in September 1938 there had already been a marked improvement in export prices and demand and, as we have seen, there was no real threat of excess production being dumped onto the domestic market. Brooke Bond's price leadership had restored profitability and there was now an administrative structure in place to allocate quotas to each producer for delivery to the Brooke Bond packing factory in Kericho.

The first official notification that East African growers had of war preparations was the circulation of the Board of Trade memorandum.[1] A Food Controller was to be appointed in the UK, who would determine the total quantities of tea required from the producing countries. Local governments would then arrange allocations, using the organisations set up to administer the Tea Restriction Scheme. The Food Controller would make contracts with individual estates for the quantities allotted to them, the tea being purchased free on board (f.o.b.), with the ministry assuming risk of shipment. It was envisaged that prices would be based on average prices during the nearest 'normal' period prior to the outbreak of war and that it would be possible to claim allowances later for increased costs of production.

In June 1939 a representative for the East Africa producers, GH Jones from one of the leading tea brokers, was appointed to the Whitehall Advisory Committee on food defence plans, and he in turn appointed three industry advisors from Brooke Bond, James

Finlay and another broker. The original plan had been that, at the outbreak of war, an annual contract would be negotiated based on the average prices for the three previous years. But it was not possible to complete the arrangements quickly enough (they required endorsement from the Ministry of Food, Board of Trade and the Treasury), and so there was an initial short-term contract to run until the end of 1939 using London auction prices in the week commencing 28 August. The three East African colonies were asked to supply 3 million lbs of tea for the contract, which applied to all tea shipped after 15 September. All tea shipped before this date, or in bond in the UK, was requisitioned by the Ministry of Food and paid for on the same terms as the short-term contract. Back in Kenya, the Director of Agriculture had been designated Tea Commissioner, and entered into contracts with producers, calling for delivery when shipping was available. The Ministry of Food agreed to make a payment on account when the tea was shipped (and before prices had been established by the brokers in London) at the rate of seven pence per pound, with the balance payable within 130 days (the conventional shipping period). This was roughly two-thirds of the final price.

The arrangements for the long-term contract of 1940 followed closely the provisions of the temporary contract, save that prices were now based on the averages received by an estate for the three years to 1938. Sales proceeds were subject to deductions for tea cess, brokerage of 1 percent, an amount of 1.28 pence per pound to represent the difference between the London sale price and f.o.b. cost, and finally interest at 5 percent on the shipping period of 130 days.[2] By this means, estates were to work out their tender price for submission to the tea controller, who in theory could pick the most favourable ones. In the event, the Ministry of Food accepted everything that was offered, so long as estates kept within their quotas under the Restriction Scheme.[3]

The contract document made provision for a price increase "in recognition of the exceptional conditions likely to obtain during the period of the contract and the possibility of increased costs of production," with a flat rate applicable to each district.[4] This created a difficulty in Kenya due to the considerable acreages of immature tea, which naturally affected production costs on those estates. The KTGA suggested retrospective payments when the increased costs could be demonstrated, but this was unacceptable to the Ministry of Food, which was required to establish costs in advance. Jones was able to strike a deal with the Director of Supplies: "We were aiming at an increase of ¾ pence per lb, as shown by the figures we presented. After discussion the various items were admitted, but it was made clear that Government were not prepared to shoulder the full amount and look to the industry to share in the burden. We finally compromised at ½ pence per lb, which we had agreed beforehand we should accept if we could not do better."[5] In his letter, Jones went on to draw attention to an important implication of the government's recognition of cost increases. "It should be noted that the allowance which has been made... is Government's contribution towards the extra costs of production due to the war. The claim for an extra allowance will have to be substantiated if a further contract is made for 1941. Producers should recognise therefore that expenditure should not be curtailed, but they should maintain their properties in the customary manner." It was no fault of the Ministry of Food that the condition of the Kericho estates deteriorated so much during the war.

The terms of the Ministry of Food contract provided that an exporter could withdraw tea in order to meet an export demand and to offer a substitute grade. A covering note explained: "The Government policy is to foster the export trade and it is possible that, owing to freight difficulties, the responsibility for this will be thrown in a larger degree than usual upon the producing countries. Not only is it important for the future prosperity of the tea industry to retain its export trade, but the need for dollar and other exchanges makes the retention of this trade during the war eminently desirable."[6] However, it was not clear to what extent producers were free to give the export trade priority over the Ministry's requirements. The latter wanted 10 million lbs for the 1940 contract; was it permissible to tender for less? In response to enquiries, the Director of Supplies wrote to Jones, "the Ministry does not wish to make purchases which will interfere with this (export) trade."[7] In Kenya this was interpreted by the tea commissioner to mean that "Contractors may reserve up to a maximum of 20 percent of their exportable quota for this purpose."[8]

Kenya's share of the 10 million lbs bulk contract was 8,615,949 lbs. Individual estates were only allowed to tender for their share as defined by the ITA quota, but were also allowed to export to other markets. The result was that there was a shortfall in the tender amounting to 834,709 lbs. When this was realised the Ministry of Food sought to fill the shortfall by offering to buy available supplies, but at a 20 percent discount on the contract price. It received no takers until it agreed to pay the full price.[9] The potential for competing demands on the Kenya tea industry, resulting in confusion in the market, was evident and the addition of two additional factors made it certain. The first was that the Ministry of Food was itself a tea exporter, through a body called the United Kingdom Commercial Corporation, the remit of which was extended to the Middle East in 1941. Second, the Middle East Supply Centre was established in Cairo in April 1941, which added another purchasing and supply organisation to the picture. The Sudan export market became alluring for the Pool, against a background of a stagnant domestic one (aside from growing military demand). Domestic tea sales by the Pool in its first year of 1.6 million lbs had been adversely affected by merchants stockpiling in anticipation of higher prices. But in its second year to September 1940 they only amounted to 1.7 million lbs, and Brooke Bond reported that native purchasing power had been affected by higher import prices and by reduced employment in coffee and sisal. Sudan had a normal consumption of 6 million lbs a year and traditionally had imported cheap grades in bulk, whereas the Kenya product was considered too expensive. However, due to wartime shipping difficulties in the Red Sea ports, the prospects had changed and Brooke Bond was keen to tackle the market. In September 1940 the Advisory Committee agreed that the Pool could sell to Sudan. This opportunity was already being seized by Uganda (which was outside the Pool), in response to an appeal from the Governor

General to increase trade between the two countries. In the three years 1939–41 tea exports from Uganda went from nil to 69,000 lbs to 253,000 lbs. The sequence for the Pool was nil, 6,000 lbs and 863,000 lbs, which was equivalent to half its domestic market sales in East Africa.

In 1941 tea production by Pool members amounted to 15.2 million lbs. The Ministry of Food originally requested 12.5 million lbs from East Africa. Military demand was picking up and the authorities contracted directly with estates, so that the Tea Commissioner advised "should the military requirements prove to be greater than can be anticipated at present, the Ministry of Food will be approached to release sufficient contract tea to meet the military demand."[10] Meanwhile domestic demand was at last responding and Pool sales rose by 30 percent to 2.1 million lbs. Finally, there was the breakthrough into the Sudan market. The strain on the control apparatus became evident towards the end of the year.

In December 1941 the Ministry of Food telegraphed that its requirement from East Africa for the 1942 contract would be 9.5 million lbs at unchanged (1940) prices, but went on to say "and we shall meet all requirements of the Sudan civil population and if necessary other Middle East Requirements."[11] It appeared as if the Ministry was about to prevent the Pool from selling in the Middle East, and had reserved 3.5 million of the proposed bulk contract for this purpose. There was the further implication that the Pool would be left with a substantial surplus of unsold tea if it was cut out of the Sudan market, since the forecast for Kenyan production for 1942 was conservatively estimated at 13 million lbs which was notionally allocated as Ministry of Food, 8 million lbs; domestic market, 2.2 million lbs; and military and local exports, 1.36 million lbs; leaving a surplus of 1.44 million lbs.[12] At the KTGA meeting in January 1942 McWilliam argued "any surplus over and above local markets and military contracts should be made available to the Sudan market through the producing agents. Also, that should it be the intention of the Ministry to divert contract teas into the Sudan, as this market has already been developed by the Pool, such teas should be handled by their agents."[13]

It appeared that the Ministry of Food intended to use the UK Commercial Corporation (UKCC) to carry out the 3.5 million lbs of tea purchases from Kenya and then to deal directly with importers in Sudan, and that any additional requirements would likewise be handed by the UKCC. Kenya protested officially at being summarily excluded from a growing export market, one for which it had developed a distribution network with tea grades and packaging, and which would be an important outlet post-war; it went on to argue that, if it was to be cut out, then the Ministry should undertake to buy all its exportable surplus tea. London then started to backtrack, asserting that there was no intention to disturb normal trade channels, and hedging its position over the UKCC. There followed a meeting of the East Africa Supplies Board in January, where it was decided that the Ministry contract should relate only to the UK and that the UKCC would purchase for Sudan, taking all available export surplus, although it was still envisaged that the Sudan government would be the contracting party, cutting out the Pool. However, these decisions were overturned in London, as the Ministry insisted on being the contracting party for the original 9.5 million lbs but no more:

> "I am now directed to inform you that the Ministry of Food
> have now advised through the Colonial Office that they do
> not wish to purchase the balance of the exportable surplus of
> East African tea from the 1942 season not already sold to
> them. Instructions have been received that the balance should
> be disposed of as follows:
> (a) exports to the Sudan will continue at the present rate only,
>   and existing arrangements will not be varied;
> (b) any balance may be exported to South Africa and the
>   Congo."[14]

In other words, the Pool would be allowed to retain its present level of trade with Sudan of 2 million lbs and be forced to find new markets for the balance of its exportable surplus. This ruling was all the odder since the letter ended by advising "I am to add that the Ministry of

Food have indicated that they will wish to purchase the entire exportable surplus of East African tea from January 1st onwards."

This ruling attempted to paper over several difficulties. A reading of telegrams shows that the War Supply Board in Khartoum had formed an opinion that a few local firms were trying to corner the tea trade. The recently arrived member of the new UKCC and the board were strongly of the opinion that all tea for the Sudan market should now be handled by the UKCC, buying direct from East Africa. The chairman of the Middle East Supply Board was converted to this viewpoint before he went to the Nairobi meeting in January 1942. Unfortunately, this overlooked the London policy that local trade should not be disturbed, which had been articulated only a few days before, and it also overlooked the fact that the tea trade with East Africa had been expressly encouraged in the first instance by the Sudan government, whose growers were now counting on continuing it into the post-war period. Indeed, the Uganda producers had gone so far as to decline to tender for the Ministry of Food contract and to send all their exports to Sudan, which was more profitable.

This was the cause of the next problem, because the Ministry of Food looked to Kenya to take up the Uganda quota. The availability of tea was not in question, but the KTGA refused: "Under no circumstances are we prepared to take up Uganda's quota as we understand their reason for not offering their quota to the Ministry of Food would appear to be guided by self interest in that they obtain a higher price for their teas in the Sudan"[15] The Tea Commissioner only had advisory powers: he proposed that the Uganda War Supplies Board should restrict export licences, but the board was reluctant to incur the negative response that would have resulted from such a move and there was a sharp exchange of correspondence. Eventually, a meeting was held between the two sides, where it was agreed that "the obvious outlet for Uganda teas is in the Sudan as it does not involve the use of the railway to the Coast."[16] Kenya undertook to fulfil the Uganda quota of 710,000 lbs to the Ministry in return for a refund of the difference between that price and the Sudan price.

Finally, the directive that Kenya should find a market for its surplus tea in South Africa came unstuck. In November, McWilliam reported that the Pool had been unable to make sales to South Africa due to its import control regulations. At this point the Ministry gave up and offered to purchase all surplus tea for shipment to the Middle East at existing contract prices and to allow the Pool to send a further 100,000 lbs to Sudan.[17]

The events of these months pointed clearly to the need for a reform of the tea control arrangements relating to East Africa. The problem had been spotlighted by one of the Uganda producers, who noted in a memorandum that "The present half-hearted and apparently uncoordinated efforts to control a part of the export surplus can only lead to trouble and muddle. If we are to have control let it be full control; otherwise let us have a perfectly free market."[18] The memorandum envisaged the appointment of a tea controller in each colony with the power to requisition tea from producers and to arrange shipments and give direction to the distributors (the Pool and its Uganda counterpart, the UTS). Fresh thinking was taking place in London as well over food control generally, and it was known that for tea the idea was to control the whole exportable surplus of the producing countries. A meeting in Nairobi in June 1942 took the London view as its starting point and recommended that the East African exportable surplus be sold to the Ministry of Food, saving a special contract for the Middle East, priced at parity with East African levels.[19] The meeting also recommended the appointment of McWilliam as Tea Controller for East Africa, with a deputy in Uganda.

In October a despatch from the Secretary of State to the East African Governors' Conference unveiled the new order: "His Majesty's Government has arranged to buy the whole exportable surplus from Empire countries. The buyer will be the Ministry of Food. Consuming countries will be given a quota by an International Committee in London, supervised by the Combined Food Board in Washington."[20] It envisaged that all East African tea production would be allocated to the Middle East and other African countries. However, as regards the views of the Nairobi meeting, "there must be a common basis of prices charged to consumer countries amongst United Nations that makes it impossible for the East African crop to command different prices in different markets. Contracts have been negotiated with India and Ceylon which have taken into account the prices previously realised by them in United Kingdom and other markets at different prices." Their contracts took the form of a compensating price increase, and a 1.5-pence-per-pound increase was eventually agreed for East Africa, following representations by the KTGA.

After consulting the KTGA, the new proposals were supported, except for a strong plea to retain regional distribution channels (the Pool): "While it is agreed that exports for adjacent territories such as Congo, Sudan, Seychelles and Somalia should be made at the new contract prices, it is considered that distribution of the approved quota to these countries should be made through existing trade channels. There are no facilities for packing and blending in the Sudan and Brooke Bond as distributors for 90 percent of the East African production may lose up to half of their trade under your proposals... they are particularly anxious that the blends which they have established as a result of efforts extending over several years should be retained on these markets. These blends include a special low grade broken tea which is not ordinarily sold to the Ministry and for which no alternative market exists. It is estimated that 1943 production will contain 500,000 lbs of this grade which would otherwise be discarded. Uganda exporters similarly desire to maintain their trade connections."[21] The Ministry accepted the argument.

It also accepted that a Tea Controller be appointed with powers to place a ceiling on East African consumption, to regulate sales to adjacent export markets and to make allotments under the Ministry of Food contracts. McWilliam was appointed to the post with effect from 1943, with a deputy from Uganda Tea Sales. The office of Tea Commissioner (the Director of Agriculture) was retained, but his functions were now limited to transmitting the decisions on the gross allocation of East African production and supervising the arrangements for the Ministry of Food contracts. Legislation was passed in all three territories to give the Tea Controller extensive powers: all tea exporters required a permit from the Controller which specified grade and packing and the price, which had to be equivalent to the ex-factory price of tea sold in the domestic market. The Controller could also require production and sales returns and any other information thought necessary.[22]

It was agreed that East African civil consumption should be limited to the total quantity of tea sold for the local trade in 1942, amounting to 3.8 million lbs, of which Pool members contributed 3 million lbs. Since there had been heavy speculative buying by traders during the year, who anticipated a sharp cut in supplies, the imposition of control was in practice not onerous. In any event, it was government policy not to interfere drastically with supplies of a locally produced commodity that was in universal demand by the African population, because imported trade goods were becoming increasingly scarce so that maintaining ample supplies of tea was a counterweight to inflationary trends.[23] The wholesale price of tea had been subject to control since the war and had been held at Shs1.40 per pound since 1937.

Under the new system, the contracts with the Ministry of Food were for the residual output of tea left after the needs of East African civil and military consumption had been met, together with those of the scheduled export territories. In 1943 the Pool tea crop was well below expectation (14.2 million lbs, as compared with 16.5 million lbs the year before), due to adverse weather conditions, and it was the Ministry's contract that was cut – from 12.5 to 6.5 million lbs, even though increased allocations were made to the War Supplies Board and exports to the scheduled territories. The tea controller had the power to ensure that the surplus after local and export requirements had been met was actually delivered to the Ministry of Food. For its 1943 contracts the Ministry allowed a price increase of one penny per pound plus an extra half a penny per pound in recognition of increased production costs (mainly imported packing materials); this represented an increase of two pence per pound since the outbreak of war, to give an average price of 13 pence per pound.

With regard to the scheduled territories, the allocations were fixed after consultation between the government concerned and the Ministry of Food (through the medium of the Colonial Office) and the amount had to be approved by the ITC. The tea controller was then informed and his function was to make specific allocations to individual producers. The Pool handled distribution for its members and others used their own trade channels. Sudan was the most important export market; here the War Supplies Board in Khartoum allocated quota import licences to merchants who dealt with the Pool, or with Uganda Tea Sales and independent Uganda growers. The Tea Controller in turn issued export licences to the growers. In this way, the quantity of tea leaving East Africa was controlled while retaining commercial channels within the overall framework. Growers were only allowed to receive the same ex-factory price as they obtained on the local market, which was a condition for the extra one pence per pound on the Ministry of Food contract.[24] However, by supplying higher grades of tea (at better prices), as the annual report of 1944 revealed "the Pool's net payment on export sales to date has been maintained at a satisfactory level compared with the Ministry's prices".[25]

The requirements of the military were notified to the Tea Controller by the East Africa War Supplies Board. These required tenders from producers and individual contracts, but in practice the Pool and the UTS were the only suppliers, which greatly simplified matters. Pricing was the same as for the Ministry of Food contract.

Up until the 1943 tea prices on domestic sales in East Africa had been approximately ten to 12 cents per pound higher than the Ministry of Food contract price, but the balance of advantage was now altered and the price for local sales was now two to three cents per pound below the Ministry's price.[26] The market in East Africa was divided into consumer zones, and the Tea Controller's task was to make an allocation to each one and to ensure regular supply over the year. The existence of the Pool greatly eased his task for, in effect, supplying the market was largely a matter of domestic Pool administration, which he was responsible for in his business capacity at Brooke Bond. In Uganda, however, there were a number of independent growers outside the Pool, which justified the existence of a deputy controller there.

The Pool agreement was due to expire in September 1944. There was little doubt that it would be renewed, although the experience of six years indicated the need for some adjustments. In Kenya, complete coverage of production had been achieved once small estates in Nandi and Sotik had joined. The same went for Tanganyika, as already noted, (including a purchase agreement for the remaining estate). Only in Uganda had the Pool failed to persuade its producers to join, and even so there was a sales agreement over prices and discounts. The outcome was that the Pool had eliminated ten brands from the market and it accounted for 81 percent of the East African domestic market. Tea consumption had doubled over the six years as Brooke Bond strengthened its distribution organisation and, by 1943, Pool sales within East Africa accounted for 21 percent of members' production. As McWilliam noted to the chairman of Brooke Bond, "this is regarded by members as a very important factor, firstly because the Pool was paying a small premium over London prices, and secondly because the effect of any security provided by a local market in the event of a collapse in world prices depends very much on the proportion of local deliveries in relation to total crop."[27]

Wartime conditions had greatly strengthened the authority of the Pool, not only in dealing with the Ministry of Food's bulk contracts, but also in the way it seized export opportunities in neighbouring markets, notably Sudan. Pool sales within East Africa in 1943 amounted to 3.1 million lbs, while exports to scheduled territories added another 2.5 million lbs and represented significant branded market development for the future. Brooke Bond itself could justifiably feel very satisfied with the Pool initiative. Its own sales organisation had come to handle the bulk of East African tea production for the domestic market and neighbouring territories, for which it received a 7.5 percent sales commission. By 1943 its sales company was earning £32,000 per year from commissions, against running costs of a quarter of that figure.

The main problem that had emerged in the working of the Pool was that the quotas were based on tea acreage and there was no recognition of the trade-off between quantity and quality. The cure was to arrange for monthly valuation samples to be assessed by London brokers. As it happened, the main complaint was in respect of Mufindi teas produced by Brooke Bond's Tanganyika Tea Company. McWilliam was flown to England in June 1944 for discussions over the renewal of the Agreement. In a memorandum at the time he floated the idea of turning the Pool into a producers' profit-sharing organisation:

"A development of the scheme which has occurred to us as being likely to appeal to the Producers and at the same time strengthen cooperation, would be to make it more properly cooperative whereby we should become partners, and the total earnings of the Pool would be divided on a profit sharing basis."[28]

However, after detailed consideration, the proposal was not put to the membership, and the Agreement was renewed substantially on its original basis.

NOTES TO CHAPTER 7

[1] MP/Sales/1/2, *Scheme for Control of Tea. Arrangements for the Purchase and Shipment of Tea from India and Ceylon.* This memorandum was sent to the KTGA by the Director of Agriculture on 19 December 1938, seeking the views of the association and who had opined "I see no particular difficulties in the Scheme for the control of tea," noting that it could be accomplished either by individual contracts with exporters or through the Pool organisation.

[2] MP/Sales/1/2, Long Term Contract document tabled to KTGA meeting on 27 November 1939.

[3] This provision added a tiresome element of red tape, since careful records had to be kept of every producer's quota, notwithstanding the general shortage of tea during the war and the fact that East African producers were nowhere near reaching their theoretical export quotas.

[4] MP/Sales/1/2, Long Term Contract 1940.

[5] MP/Sales/1/2, Letter to KTGA, 22 March 1940.

[6] MP/Sales/1/2, Covering Note to Long Term Contract.

[7] MP/Sales/1/2, circulated with KTGA letter to members, 26 January 1940.

[8] MP/Sales/1/2, quoted in KTGA circular letter, 7 March 1940.

[9] MP/Sales/1/2, Tea Commissioner cables to KTGA, 21 September 1940 and 21 November 1940.

[10] MP/Sales/1/2, Tea Commissioner to KTGA, 30 January 1941.

[11] MP/Sales/1/2, KTGA minutes, 2 January and 27 January 1942.

[12] Actual Kenyan production that year turned out to be 17.68 million lbs, so the problem was potentially much more serious.

[13] MP/Sales/1/2, KTGA minutes, 2 January 1942.

[14] MP/Sales/1/2, Letter from East Africa Supplies Board to the KTGA, 1 October 1942.

[15] MP/Sales/1/2, resolution, 2 February 1942.

[16] MP/Sales/1/2, KTGA Minutes, 24 March 1942.

[17] MP/Sales/1/2, KTGA minutes, 17 November 1942. At this meeting the KTGA, noting the work done by him in representing the producers, "unanimously voted its appreciation of the work done by Mr McWilliam on behalf of the KTGA." He had been elected an honorary member of KTGA at its annual meeting in February.

[18] MP/Sales/1/2, Memorandum entitled 'Export Control', by Sabin of Buchanans, 24 March 1942.

[19] MP/Sales/1/2, Meeting held on 3 June 1942.

[20] MP/Sales/1/2, Dispatch No. 69 of 1942.

[21] MP/Sales/1/2, Cable to Colonial Office, 13 November 1942.

[22] Kenya Government notice 619, 'The Defence (Control of Export of Tea) Order', 1943.

[23] MP/Sales/1/3, 1943 Pool report, 18 October 1943.

[24] Where the Controller could not obtain ex-factory price data, the procedure was to make a deduction from East African wholesale prices of excise duty and wholesale commission and make an allowance for transport costs to arrive at an f.o.b. price.

[25] MP/Sales/1/2, 1944 Pool annual report, 19 October 1944.

[26] Data from Brooke Bond's Kericho estates give ex-factory prices on the Ministry of Food contracts as 78 cents per lb in 1940 and 95 cents per lb in 1943. The local wholesale price of 140 cents per lb, adjusted back to the ex-factory price (McWilliam's memorandum of 18 March 1944), gives a comparison figure of 92 cents per lb for the domestic market. The Pool reported that the average prices realised for its members in the year to May 1944 were as follows: E.A. market, 91.1 cents; military contracts, 95.3 cents; and exports to Scheduled Territories, 94.4 cents. MP/Sales/1/4, letter to Uganda Tea sales, 5 December 1944).

[27] MP/Sales/1/3, McWilliam wrote a lengthy note, 'Memorandum on the Tea Distribution Agreement 1938-44', on 18 March 1944 to the Brooke Bond chairman.

[28] MP/Sales/1/3.

8

POLITICS AND PROFIT

The late 1930s saw major developments in the marketing of Kenya tea, both domestically and for export. The formation of the Pool in 1938 was closely followed by bulk purchase and allocation arrangements triggered by the outbreak of war. The creation of the role of Tea Controller, who had the combined responsibility of chief negotiator for the tea industry and the implementation of production and export quotas, led by 1946 to tea being a highly regulated sector of the economy. Producers were given precise quotas for supplying the domestic market and for exports, backed by legislative authority; meanwhile, export prices were negotiated centrally and prices for the domestic market (established by Brooke Bond in 1937) were reinforced by official price control. Brooke Bond handled all distribution for Pool members in East Africa and for the scheduled territory export markets, so that tea estates only had to deliver their manufactured tea either to the packing factory or to Mombasa for the Ministry of Food contracts. From 1946 this regime was subject to increasing strains as the instruments for orderly allocation of supplies came to be seen by the industry as instruments of oppression, and as the industry became caught up in official attempts to control the rising cost of living. It was a paradox that the success of Brooke Bond in developing a comprehensive system of retail distribution in the townships and rural areas of East Africa, coupled with the taste for tea acquired by returning askaris (African soldiers) from war service, led to rising consumer demand. A further complicating factor was that the governments of Kenya, Uganda and Tanganyika had differing views on price controls as a policy instrument.

But first, there was a reward for patient diplomacy: the three producer members of Uganda Tea Sales joined the Pool in March 1945. The Pool had formally approached the UTS in the previous year to reconsider membership and this led to a meeting with McWilliam, who explained the revisions to the Pool agreement that had been agreed as regards valuations and quotas. At this point the UTS members were still exercised over the higher distribution costs of the Pool and at the prospective loss of their trademarks.[1] It was conceded that UTS could retain its trademarks until the end of 1946, in order to use up stocks of labels, and the members agreed to join the Pool on 1 March 1945. This raised the Pool's coverage of the East African planted acreage to 91 percent, and McWilliam was congratulated at the annual general meeting "on the success of his untiring efforts to bring the producers of all three territories into the cooperative distribution scheme."[2]

Already in 1945 the tea quota allocated to the domestic market had fallen short of demand and the Pool report for the year noted that, notwithstanding a 15 percent increased allocation of 583,000 lbs, "despite a carefully worked out system introduced by the Distributors for registering and allotting individual monthly quotas to retail traders, combined with the elimination of wholesalers, black marketing practices became general... and the demand for tea everywhere by the African was increasing out of all proportion to the supplies available in the market... the wily Asian trader has been quick to appreciate the value of stocks of tea in relation to the rest of his trade, with the result that the limited supplies go under the counter to be kept for favoured customers and to be used for pushing sales of other goods."[3]

In 1946, London agreed that a further 1.1 million lbs be made available for local consumption. The Ministry of Food, the

Secretary of State explained, "do not consider that in post-war conditions it would be practicable to restrict East African consumption below what you consider to be a necessary level."[4] The Ministry of Food conceded a one-penny increase on its purchases, giving an average ex-factory price of Shs1.01 per pound. The Pool negotiated a small price increase on its scheduled territories exports which, together with the benefit of shipping higher-value grades, gave the same net return. These achievements helped to offset the opportunity cost of increased sales in the local market, where the Pool accounted for nearly 95 percent of East African consumption at prices still based on 1937, which equated to an ex-factory price of 91.5 cents per pound.[5]

1947 marked the first substantial revisions to the wartime control system. An inter-territorial meeting on 9 January (which also recommended East Africa's withdrawal from the ITA) came to the conclusion that restrictions on exports to countries other than the UK should be removed. The decision was summarised in a cable to the Colonial Office:

"Tea exports from East Africa henceforward will be free from the price, destination and quota restrictions hitherto enforced... A quantitative restriction on the total exports of tea from British East Africa however will remain in order to safeguard East African internal requirements, and permission to export tea will be confined to manufacturers and manu-facturer's accredited agents."[6]

The effect of this decision was to enable East African producers to obtain world prices for their tea in export markets for the first time since the war. Since there was now no control over export destinations, the Ministry of Food had to compete in the open market for tea. India and Ceylon had already refused to continue with bulk contracts and forced the Ministry to pay market prices following the reopening of the Calcutta and Colombo tea auctions (the London market remained closed). The Ministry would have liked to secure the East African crop with a bulk contract, and it offered an increase of four pence per pound over the 1946 prices. However, this only offered producers an average price of 17.3 pence f.o.b., whereas export sales in 1947 averaged 27 pence per pound, and McWilliam warned, "It is unlikely in my opinion that any material quantity of tea will be made available from East Africa to the Ministry this year."[7] In the event, only 11 percent of the exportable surplus was sold to the Ministry in 1947.

The purpose of tea control was transformed by this new situation. Its duty now was to apportion internal supplies equitably, so that all could participate in the lucrative export market. Manufacturers were allotted a percentage of their production to be retained for local consumption, but provision was made for the exchange of export licences between manufacturers, so that the administration could be as flexible as possible. This procedure worked smoothly in Kenya and Tanganyika, where all the factories were members of the Pool, and Brooke Bond was able to guarantee to meet market requirements and make necessary adjustments between Pool members. In Uganda, the situation was very different. About one-third of producers had not joined the Pool. They were interested primarily in the export market and had not attempted to compete strongly with the Brooke Bond brands in the local market. The tea control administration in Uganda failed to perform and licences were issued by the government to the full extent of available stocks; meanwhile, Pool members were having to reserve 39 percent of production for the local market, rising to 45 percent in 1948. With 100 cents a pound difference between domestic and export prices, considerable ill feeling developed amongst Pool members who were meeting the Uganda shortfall at an opportunity cost to themselves. McWilliam produced a paper proposing an East Africa-wide licensing authority and provision for trading export rights, but its recommendations were not taken up and the problem was eventually resolved by the progressive alignment of domestic and export prices.[8]

Tea producers became restive at the pegging of the price they received in the domestic market at the level established ten years before, in 1937. The Ministry of Food had been prepared to recognise increased production costs in its bulk contracts during the war, and now the open-market price for exports had risen markedly. The political problem for the East African governments was that tea was a constituent of the cost of living for African wage earners,[9] and also public opinion (especially with regard to Kenya settlers) was both vocal and influential on cost-of-living matters. The tea industry was perceived to be prosperous and there was little official compunction in expecting it to subsidise the cost of living. The KTGA succeeded in having its case reviewed in January 1947, seeking an increase of 20 cents a pound, or 14.5 percent, and deployed the argument "that it was not their wish to try and obtain parity with export prices. On the contrary, that the long term policy of fostering the local trade with the object of further expanding consumption for the future benefit of the industry, remained as a principal objective, notwithstanding the attraction of high prices which were likely to rule on a free export market."[10] Although the review committee agreed that there was a compelling case for an upward adjustment, the request was rejected "due to the marked effect it would have on the cost of living."[11] However, further representations led to a retreat a month later and the 20 cents award was conceded.

There was a lull during 1948 until the Kenya growers reopened discussions with their government in November. It was felt that Kenya should not act alone and so an inter-territorial meeting was convened in March 1949. The outcome was to set up a body called the Interim Central Tea Committee to secure an East African solution on tea prices.[12] In Kenya, the Executive Council then remitted the problem to the Cost of Living Commission that was sitting at the time. Although this body was primarily concerned with protecting the interests of consumers, it recognised the case for a further increase in tea prices due to higher production costs (which had risen by 75 cents a pound since 1937), and recommended a flat increase of 25 cents. This was accepted by the government in June, but the impact on consumers was ameliorated by a decision to abolish the 15 cent excise duty, on condition that the benefit was passed on. This was implemented in October and was followed by Uganda and Tanganyika.[13]

It was accepted policy that tea prices within the Customs Union should be uniform, and that any adjustments would have to be justified on the basis of cost of production considerations. This was a problem for Tanganyika, whose production costs were much higher than those of the other two territories (and 50 cents a pound above the approved wholesale price), but whose teas commanded a premium price in the export market, so there was resentment at having to reserve a substantial proportion of production for the local market. On the other hand, a price that recognised Tanganyika costs could not be justified in Kenya or Uganda. Producers in Uganda favoured moving to market prices and abolishing price control, while Kenya producers wanted to find a solution at 'a fair price' dissociated from export values, but better than the award by the Cost of Living Commission. An atmosphere of crisis began to develop towards the end of 1949 precipitated by Tanganyika, but in all three territories opinion was hardening against continued price control.

A special meeting of the East African Production and Supply Council was convened early in November 1949, for which both the Kenya and Uganda producers submitted memoranda arguing for the removal of price control.[14] The Uganda paper concentrated on a moral argument: between 1939 and 1949 the producer price had been raised by 45 cents a pound, whereas the Ministry of Food had awarded increases totalling 98 cents. The industry was now subsidising the cost of living at the rate of £300,000 a year. "Can it be argued that it is in any way fair that an industry having a local market should be thus penalised as compared with an industry having little or no local market?" the paper asked. "Would it not be more equitable for any such subsidy to be shared equally over the whole tax paying community by way of a food subsidy than by this method?" The Uganda Director of Agriculture was sympathetic to the notion of moving to world market values as opposed to prices based on cost of production and an approved margin. In a memorandum he argued that "the general trend is in this direction. Producers of butter in Kenya now receive a price for all sales, which is directly linked with the New Zealand price. Coffee growers receive world prices for their production, the Uganda cotton crop is sold at the full overseas value as is also East African sisal. In Uganda it is only with such crops as tea, sugar, oilseeds and tobacco, where a high proportion of total production is required for local consumption that producers are penalised in that prices on the local market are kept by direct Government action at figures less than export values."[15]

The Kenya paper reviewed the history of price control during the war and its lack of recognition of rising costs or of the growing proportion of output being reserved for the domestic market, which had reached 45 percent in 1949, and went on to point out that the situation was discouraging new investment. Its recommendation was for an annual price review related to export values. In return for this, the Pool would guarantee normal supplies to the market.

At the November meeting, McWilliam explained the difficulties of achieving a uniform solution for the region.[16] Production costs in Tanganyika were above the control price for the domestic market; Uganda had non-Pool growers unable to sell their local market allocation; and Kenya had found itself maintaining supplies to the whole East African market. Some form of price equalisation was needed, he argued. The tea industry was instructed to come up with a workable scheme for securing supplies to the domestic market, with an appropriate regional adjustment to the producer price, in the knowledge that both Kenya and Uganda wished to retain both price control and export licensing. The Interim Central Tea Committee was empowered to approve a solution, in the knowledge that failure to do so would result in the Tanganyika Government taking unilateral action.

The KTGA met in preparation for this meeting and formulated proposals for an East Africa-wide price increase of 30 cents per pound and for an equalisation amount of ten cents per pound to finance an adjustment between export and local prices, in order to facilitate the transfer of export rights between Pool members and for the payment of valuation premiums.[17] In return, Kenya growers would undertake to maintain supplies to the East African market and to hold prices stable for a year. After much debate these proposals were adopted by the Interim Central Tea Committee in December 1949, but effective only until the following June.[18]

By early 1950 it had become an open topic that the era of price control was coming to an end. The Tanganyika Government had given notice of independent action to lift both export and price controls if an agreed solution was not achieved.[19] The Uganda Government still wanted to retain price control, while the Kenya Government did not wish to see another price rise before the end of the year. There was uncertainty as to the effect on prices of the removal of control, but Brooke Bond advised that there would be smuggling if there were differential prices within the Customs Union. In the event, the Pool agreed to continue its guarantee arrangements for another three months while the newly established statutory Tea Boards aimed to reach agreement on a new East Africa price level, advised by the Pool.[20] The Pool wished to see another uplift in the controlled price of 50 cents, but this was opposed by both the Uganda and Kenya governments. The weak link in the system was high-cost Tanganyika and the Pool was able to force a 100 cent price increase there in December 1950, under the threat of cutting off reciprocal arrangements to allow Tanganyika to export all its tea and for its local market replacement to come from Kenya. The Uganda Government was the next to come to terms.[21] Meanwhile, in Kenya, McWilliam wrote a lengthy paper for the Tea Board reviewing price control and the operation of the Pool since 1945, highlighting the anomaly of price control without rationing and the associated increase in local tea consumption, and making the case for a 100 cent price increase: "In the UK the price is subsidised by the general taxpayer whose liability is limited by the rationing system. In East Africa the retail price is subsidised by the produce at a very low level and there is no limit to his liability."[22] The government agreed and, in return, the Tea Board offered an undertaking to hold prices for 12 months and also that the Pool would fully satisfy the local market and would maintain the local price level below the export parity price.[23] On these terms, price control was suspended in October 1951.

Despite the suspension, there was a sting in the arrangements: a fall in export prices would bring into play the Pool's undertaking to keep domestic prices below export prices. Furthermore, the Kenya Government still expected to be consulted by the Kenya Tea Board on price changes, and to have an influential say in the outcome. World market prices fell during 1952 and, in August, the Pool made a ten cent reduction in the wholesale price. The Tea Board felt it necessary to warn government that its influence (via the Pool) stopped at the wholesale price level: "short of stopping supplies altogether, we cannot exercise any control over the retailer's selling price."[24] Nevertheless, negotiations took place between the Tea Boards and the three governments over the pricing policy for 1953, which resulted in the anodyne conclusion that wholesale prices be unchanged. However, auction prices commenced a sustained rise at the beginning of 1953, which led to restiveness amongst producers whilst their representatives again consulted with government. The Pool Advisory Committee passed a resolution in September "to abolish any form of price control on tea,"[25] but the Kenya Government wanted to establish a regime of an annual price review, based on the previous year's average prices. The Pool acceded and announced a price increase of 68 cents in January 1954 to Shs3.05 per lb.[26] With auction prices continuing to rise during the year, there was strong pressure for a large increase in 1955. In November 1954 the Pool resolved to press for full discretion to make price changes and the government retreated, agreeing to take no further part in setting the tea price. The Pool responded by deciding on a 59 percent hike to Shs6.12 per lb in the wholesale tea price.[27]

This was the end of official intervention in tea prices in East Africa. During the period of price control East African tea consumption had increased by 73 percent over the five years to 1950 to 7.5 mn/lbs. However, the Pool had overreached and consumer resistance was experienced with consumption fluctuating around 8.2 mn/lbs for six years and it was the Pool's turn to retreat. It cut prices by 62 cents in April 1955 to a weighted average of Shs5.50 per lb, and then by a further 100 cents in July, against audible threats from officialdom to introduce an export tax.[28] Sales of 9 mn/lbs were finally achieved in 1957.[29] In freeing itself from 15 years of officially administered prices, the tea industry had submitted itself instead to world market pricing. Nevertheless, this period of control and interaction with officialdom had left its mark.

The Pool was conceived as a cartel in the 1930s in order to overcome cut-throat competition in the domestic market at a time when new planting was coming into being and when exports were to be restricted under the ITA. The outbreak of war, with its Ministry of Food bulk contracts and controlled prices, greatly enhanced the authority of the Pool and its attendant Brooke Bond marketing organisation. As we have seen, the Pool developed an export dimension to Sudan and other scheduled territories in Africa; however, its most important impact was within East Africa, and especially within Kenya. The legacy was to establish tea drinking as a basic consumer habit amongst the local population, through comprehensive distribution, and this occurred with increasing success as tea provided a healthy and safe alternative to polluted water or beer. Brooke Bond set about creating such a system, derived from its experience in Lancashire and India. To begin with, tea was sold through a network of Indian agents and wholesalers but, from 1945, Brooke Bond moved progressively to dealing directly with individual retail outlets around the country. This entailed setting up a network of depots in the larger towns throughout East Africa, from which retailers were supplied by handcart deliveries in the towns, while over-the-counter sales were made to outlying retailers. This was superseded from 1950 by Brooke Bond's classic system of direct sales to shops throughout the region, delivered by branded vehicles in the colours of its principal brand of *Simba Chai*. This intensive coverage resulted in more than 15,000 retail visits a week.[30]

"Gradually we began to build up a pattern of the best way of introducing tea to rural populations," Hugh Morton, Brooke Bond's former sales director in Kenya, remembers. "We had several vans fitted up as mobile kitchens with loud speakers and painted in our colours. We went out on market days driving the vans into the middle of the market place. We would turn on the loud speakers and give them some music and then a pep talk. Then we would distribute cups of tea to show them the correct way of preparing tea. We wanted to prove that it was a good drink, it did them good, and wasn't expensive... Our main competition came from Coca-Cola and cigarettes, who were very good marketeers. It was always a battle with them to get the available cash in the market place. We did it, really, through the regularity of our calls."[31]

In this way the shopkeeper, instead of being forced to purchase supplies from the nearest market town, was able to receive direct deliveries as cash sales, in quantities and packages matching previous business. The aim here was to ensure fresh tea on the shelves; to avoid stock speculation; and to follow the official price levels. Nevertheless, there were disruptions to this arrangement. On the coast there was a seasonal smuggling trade during the 'dhow season' since merchants could obtain much higher tea prices there. (During the season, one monsoon wind blew the Arab sailing vessels down from the Persian Gulf to East Africa, loaded with carpets; the next allowed them to return with a cargo of tea or, in former times, slaves.) For a short time a pattern of smuggling developed with Tanganyika when the controlled price of tea was raised in November 1950, but not in Kenya: "This had the almost immediate effect of decreasing sales at our Tanganyika depots in those towns near the borders of Kenya and Uganda as dealers in those areas found that there was nothing legally to prevent them from purchasing tea outside Tanganyika at current Kenya and Uganda retail prices."[32]

Did price control retard the post-war expansion of the Kenya tea industry? This was a widely held view within the industry. It was reflected in McWilliam's paper of October 1949, which claimed that "the Government's price control policy is regarded as a strong discouragement to any large scale expansion of the tea industry by reason that the combination of low prices and increasing local consumption holds out no prospect of a satisfactory return from the capital invested."[33] It also influenced Cavendish-Bentinck, Member for Agriculture, who wrote to the Tea Board that he was "perfectly satisfied that in the interests of development of the tea industry a price increase is justified and that he would continue to do everything possible to assist the growers in their efforts to get an increase."[34] When price control was formally abolished in 1951, *The Times'* correspondent filed a despatch supporting the argument of inhibited development: "The Government declares that it has ample evidence of a falling off of interest in new tea development in the colony during the past year, and the decision has been taken in the interests of longer term development of an important industry. It is regarded as significant that of approximately 54,000 acres licensed for tea development in Kenya only 18,000 acres have been planted."[35] Villiers-Stuart of James Finlay annotated the article, suggesting that the disparity between licences and development was influenced by farmers in borderline areas for tea growing hoping to boost land values by obtaining a licence. Nevertheless, a degree of scepticism appears warranted. Between 1947 and 1952 the Kenya tea acreage expanded by a quarter to 21,000 acres, representing an investment of some £900,000 at the cost of £200 an acre.[36] Given manpower and other shortages at the time, it is doubtful whether a much faster rate of development could have been achieved. Production costs were below 60 cts/lb during the war but rose sharply in the period up to 1951; however there was still a satisfactory margin on Pool sales. With the mature pre-war plantings undertaken at a capital cost of around £72 an acre, Kericho's 12,000 acres of pre-war planted tea was yielding a post-tax return on capital of over 20 percent, and shortly much more as the benefit of higher world prices flowed through.[37]

As we have seen, it was Pool policy to meet local tea demand as a first call on production. Post-war sales in the East African market grew strongly: from 4.3 million lbs in 1945 to 7.5 million lbs in 1950 and on a still rising trend. It was this pre-emption of export revenue that was a prime source of concern among members of the Pool. It meant that, for much of the period, over 40 percent of production was directed to the domestic market instead of to exports. A rough estimate of the opportunity cost in terms of foregone revenue over the eight years suggests a figure of around £1 million.[38] These foregone earnings were of course a significant reduction for the industry, although they do not appear to have inhibited a robust planting programme. It may also be observed that another consequence of depressing the industry's earnings through the cost-of-living subsidy was to reduce government tax revenues.

During the years of licensed exports, bulk contracts and controlled prices it clearly suited government to have the clear channels of communication and execution that were provided by the Pool and the KTGA. It also suited Brooke Bond in consolidating the authority of the Pool, and McWilliam gave effective expression to this approach: doubling up as tea controller and manager of the Pool, acting as the diplomatic go-between when differences arose and preparing key papers. The establishment of the Tea Board of Kenya in 1950 gave formal expression to the linkage of public and private sectors. The Tea Board, with its membership drawn from producers and government, had statutory powers to licence tea growing and to represent the industry.

An early example of the government's tendency to view economic development in corporatist terms was seen in early 1952, when the Minister for Commerce and Industry, as a board member, presented a paper calling on the Tea Board to frame an export policy for the industry, set up a market-research capability and even to handle tea sales on behalf of producers.[39] This was a step too far for the tea companies, who feared a costly burden on the industry: "Nothing which the Tea Board might be able to offer in the way of marketing facilities would be likely to replace the recognised outlets which have become established over many years and are constantly being developed."[40] Instead, attention would be better directed to improving the quality of Kenya tea. And, indeed, the Tea Board took over the running of the Tea Research Institute that had been started by Brooke Bond.

After independence, the network of Indian-owned rural *dukas* (trading posts) in the rural areas disappeared and Brooke Bond's direct marketing network came under political challenge from the emergent class of Kenyan wholesalers. In 1974, Brooke Bond retreated and closed its network in favour of supplying wholesalers. Another source of tension arose as a result of government exercising control once again over the wholesale price set by the

Pool, which disconnected domestic prices from those obtained through exports. This led to restiveness amongst producers and a reluctance to see growth in the domestic market; it also led to smuggling to Tanzania and Uganda. Matters came to a climax in 1977 when high export prices led to resignations from the Pool cartel and widespread discontent with its management. In 1978 the government stepped in, removing the management of the Pool from Brooke Bond and setting up a new entity, the Kenya Tea Packers Association. Initially, the smallholder tea authority, the Kenya Tea Development Authority (KTDA), was tasked with running the new enterprise until government decided that it should be independently managed. Nevertheless, ownership of the Kenya Tea Packers Association, or Ketepa as it became known, was allocated according to the acreage of contributing producers, with the result that KTDA was the dominant 67 percent shareholder and provided the initial chairman. Ketepa purchased Brooke Bond's packing factory in Kericho and effectively stepped into its predecessor's shoes, levying quotas on producers and still being subject to official price control.

In contrast to the Pool's historic role of promoting tea consumption in Uganda and Tanzania and other African markets, especially Sudan and Egypt, Ketepa was confined to Kenya and disbarred from exporting. It was not until 1992, as part of the liberalisation policies advocated by the World Bank, that government ceased to exercise control over the domestic price of tea. Instead of assessing quotas from producers, Ketepa henceforth purchased its requirements at Mombasa auction prices, which removed a major source of tension with producers. The reforms also meant that rival tea packers could enter the domestic market, and they did so aggressively. From supplying virtually 100 percent of domestic consumption, Ketepa's share fell progressively over the following years, reaching around 60 percent of the market by the end of the century, with little overall growth.[41] As a result of liberalisation the organisation was encouraged to become an exporter again, especially of packaged teas, and this has become a key priority. Ketepa is now an acknowledged subsidiary of KTDA (whose shareholding has increased to 80 percent), and has an important role in promoting awareness and sales of Kenya teas. However, it has not sought to replicate the intensive retail distribution methods pioneered by Brooke Bond that played such an important part making Kenya a tea-drinking society.

NOTES TO CHAPTER 8

[1] MP/Sales/1/4, Note of discussion between the UTS and McWilliam, 21 August 1944.

[2] MP/Sales/2/1, Pool Annual General Meeting, 6 March 1945.

[3] MP/Sales/2/1, 1945 Pool report. The Tea Controller's records show an even larger allocation of 1.1–4.3 million lbs.

[4] MP/Sales/2/2, Secretary of state to secretary Governors' Conference, 9 October 1946.

[5] Pool sales in East Africa amounted to 5.3 million lbs; Ministry of Food contracts and exports to scheduled territories amounted to 10 million lbs.

[6] MP/Sales/2/2, Secretary Governor's Conference to Secretary of State, 20 January 1947.

[7] MP/Sales/2/2, Tea Controller to Director of Produce Disposal, 15 March 1947.

[8] MP/Sales/2/2, 'Establishment of a Central E.A. Authority for Licensing Tea Exports', 11 January 1949. The paper estimated that the Pool had had to make good 541,000 lbs in 1948, at an opportunity cost to members of £27,000.

[9] The statutory minimum wage formula in Kenya included half a pound of tea a month in the ration scale.

[10] MP/Sales/2/2, Tea Controller to secretary to Governor's Conference, 14 January 1947.

[11] MP/Sales/2/2, Director of Produce Disposal to Tea Controller, 14 February 1947.

[12] Its other duty was to coordinate projected tea legislation in the three territories to set up statutory boards to regulate the industry.

[13] On 1950 production, the revenue sacrifice was £112,000 in Kenya, £28,000 in Uganda and £13,000 in Tanganyika.

[14] MP/Sales/2/2, Uganda Tea Association, 'Case for Removal of Local Price and Export Control and/or Increase in Domestic Price of Tea', TT Simpson, 10 September 1949. 'The Kenya Case for the Removal of Price Control', D S McWilliam, 24 October 1949.

[15] MP/Library/25, Memorandum, Director of Agriculture, 26 September 1949. Thesis, p.276, where there is a more detailed account of the negotiations

[16] MP/Sales/2/2, Meeting of EA Production & Supply Council, 15 November 1949.

[17] MP/Sales/2/2, KTGA minutes, 25 November 1949, which also give a concise summary of the Production & Supply Board conclusions.

[18] MP/Sales/2/2, ICTC minutes, 8 December 1949.

[19] MP/Sales/2/2, Chief secretary, Tanganyika to E.A High Commission, 12 May 1950.

[20] MP/Sales/2/2, ICTC minutes, 30 May 1950.

[21] MP/Sales/2/2, Uganda Standing Economic Committee, 5 June 1951.

[22] MP/Sales/2/2, East African Tea Prices, 31 May 1951.

[23] MP/Sales/2/2, Summarised in Chief Secretary Kenya to his opposite number in Uganda, 28 September 1951.

[24] MP/Sales/2/2., Kenya Tea Board to Chief Secretary, 19 March 1952.

[25] MP/Sales/2/3, Advisory Committee minutes, 22 September 1953.

[26] MP/Sales/2/3, Advisory Committee minutes, 18 December 1953.

[27] MP/Sales/2/3, Advisory Committee minutes, 30 December 1954.

[28] MP/Sales/2/3, Advisory Committee minutes, 22 March 1955. Also, MP/Sales/3/1 Tea Price List 1954–58.

[29] MP/Sales/3/1, East Africa Pool sales: 000s lbs: 1951 – 8,336; 1952 – 7,754; 1953 – 8,851; 1954 – 8,630; 1955 – 7,108; 1956 – 8,709; 1957 – 9,007.

[30] MP/Sales/3/2, The Sales Organisation in East Africa. *The Tea Chest*, 1956.

[31] MP/Sales/3/2, Descriptive article in the Brooke Bond house magazine, *The Tea Chest*, 1956 and a recorded memoir by Hugh Morton, sales director MP/IM/1/9 in December 1997. Kenyan-born, Morton spent six months in India in 1947 visiting Brooke Bond's distribution system in India before applying it to Kenya.

[32] MP/Sales/2/3, Brooke Bond sales report, July/December 1950.

[33] MP/Sales/2/2, 'The Kenya Case, op. cit.

[34] MP/Sales/2/2., Letter to Kenya Tea Board, 12 October 1950.

[35] *The Times*, 25 October 1951.

[36] MP/Sales/2/2, Letter from Chairman Kenya Tea Association to Director of Agriculture, 7 December 1951. The cost of pre-war initial planting was put at £72 an acre. These figures may not have included factory construction costs, as no details were provided.

[37] MP/BB/1/2, Kenya Tea Company figures 1925-56 provided by the Finance Director, March 1957.

[38]

	E.A. Pool Sales '000lbs	Pool Price cents/lb	Export Price cents/lb	Opportunity Cost £
1945	4,323	90	103	21,615
1946	5,323	91	110	50,230
1947	5,017	99	173	185,629
1948	6,123	109	154	140,829
1949	6,889	119	197	272,155
1950	7,440	155	241	329,000

Prices from MP/Sales/2/1 and 2/2, Pool Sales in MP/Library/25, Thesis, p.337 table.

[39] MP/Sales/2/2, *Marketing Prospects for Tea*, Arthur Hope-Jones, 18 April 1952.

[40] MP/Sales/2/2, *Marketing Policy*, Tea Board paper, 11 July 1952.

MP/IM/2/8, Interview with Ketepa managing director, 5 September 2003.

9

DEVELOPMENTS IN
THE ESTATES SECTOR

In anticipation of the ending of the Tea Restriction Scheme, the Kenya Government began preparing for a big expansion of the tea sector by making available additional Crown land for plantation development. Its aim was to attract investment from tea companies with estates in India and Ceylon, where political developments in these countries had stimulated contingency plans as they approached independence. In this it was successful and two groups in particular made important commitments to Kenya over the succeeding years. Meanwhile, both Brooke Bond and James Finlay prepared to greatly expand their existing interests.

The government's plans for tea entailed selling Crown land at Kimulot, which was adjacent to James Finlay's existing estates, and at Changoi, as well as excising part of the Mau forest at Timbilil, which was of interest to Brooke Bond for the same reason. The government also received a controversial proposal from Brooke Bond to convert its long lease from the Kipsigis' native authority, arranged in the 1920s, of the land forming its first estates at Chebown-Kerenga into freehold.[1] This application, if granted, would result in a permanent reduction in the area that had been designated as the Kipsigis land unit and it aroused strong opposition from within the administration, as well as from the native authority.

In August 1947 the Chief Secretary informed the Nyanza Provincial Commissioner that Executive Council had agreed to the granting of freehold to Brooke Bond. In return there was to be a compensatory grant to the Kipsigis of Crown land (its extent as yet undecided) at Kimulot.[2] Some months earlier he had opined "Kimulot represents some of the only remaining first class land in the Colony and applications have already been received for it to be alienated for that purpose... and it is considered that it should be dealt with by the European Settlement Board for closer settlement and the cultivation of tea."[3] The District Commissioner in Kericho, Tony Swann, was alarmed and wrote to his Provincial Commissioner: "I think there is a very real danger of severe political repercussions if Kimulot is alienated in its entirety, as the renting of Kerenga and Chebown from the Native land Unit is, as far as I know, unprecedented in Kenya."[4]

The linking of Kimulot to the freehold deal proved highly contentious, as the Kipsigis regarded it as their territory in the first place. Unsurprisingly, the Native Council minute recorded the opinion that "the whole of Kimulot should be granted to the Kipsigis. They were not willing to receive less as they were originally in occupation of all this land before removal by government... Chebown-Kerenga should continue to be held on lease to the company."[5]

The government first proposed to excise 4,500 acres out of the 12,000-acre Kimulot area to the Kipsigis in exchange for the loss of the 6,500-acre Brooke Bond lease. Under pressure, the offer was increased to 5,000 acres plus the lease of two government-owned farms in Sotik. In September 1948 it was raised again to 6,500 acres, but Swann criticised the offer, restating his view that the Kipsigis should be offered the whole of Kimulot in settlement of the exchange, since the government was also proposing to excise another portion of the reserve at Changoi: "In view of the fact that all the land in question was formerly part of the integral tribal lands of the Kipsigis, I do not think that such a proposal could be decided as unfair."[6] In rejecting this view, the Member for African affairs, Percy Wynn-Harris, then threatened to alienate the whole of

Kimulot to tea companies on 999-year reversionary leases to the Kipsigis if the Native Authority refused "the final solution to the Kipsigis problem" of an acre for an acre.[7]

However, there was yet another iteration in November 1949 when the Council of Ministers, the Highlands Board and the European elected members all agreed to increase the offer to 7,250 acres and this was accepted by the Native Council in December 1949. A year later, the Provincial Commissioner Nyanza received a letter from a member of Legislative Council, BA Ohanga, stating that the decision by the Local Native Council "did not represent the opinion of the Kipsigis indigenous elders, whose opposition is so strong."[8]

Having had one portion of their traditional lands leased to Brooke Bond for 999 years on their behalf by the government in the 1920s, a loss now made permanent by its conversion to freehold, the Kipsigis were having restored to them as compensation some 60 percent of Kimulot, which they had always claimed as their territory. This area, between the Koruma and Itare rivers, was then designated an official settlement scheme and subject to strict controls; it was divided into 214 farms and some of the first African-grown tea was planted there in 1955.

After the war, the tea companies became aware that the government was proposing to excise tracts of Crown land in the Mau Forest for European settlement at Sitoten and Tinet. On learning that another 12,000 acres were to be excised at Timbilil, Brooke Bond expressed interest. The Forest Department strongly opposed this change of use and it was supported by the Forest Boundary Commission on several grounds: that 15,000 acres had already been excised for European settlement; the suitability of the area for tea was unproven; apprehension that there would be adverse climate effects from the removal of forest cover; and – more self-interestedly – the lost opportunity this presented for softwood forestry development. Even the KTGA opposed the change of use. The dispute between officials and the settler-dominated Highlands Board delayed matters for more than five years, and during this time the government hydrologist reported that there would be no adverse effects on river flows, which led to a referral for further research. Eventually, in 1955, the Council of Ministers insisted on the forest excision but stipulated that it remain under the control of the Forest Department until the research was finalised. In May 1957, tenders were advertised for 1,500 acres and Brooke Bond secured the land at £7 an acre.[9]

Brooke Bond's tea acreage had been frozen at 3,220 acres in 1933 by the Tea Restriction Scheme, so far as new planting was concerned. However, during the war it was able to purchase three small estates in Kericho: Kapkorech (407 acres) from Lord Egerton; Chelimo (269 acres) from Caddick; and Karabwet (87 acres) from Orchardson. Shortly afterwards it purchased the rest of the Egerton tea at Jamji (600 acres), thus raising the total acreage to 4,583 acres.[10] With Kenya's withdrawal from the post-war renewal of the Restriction Scheme, Brooke Bond embarked on a major new planting programme on its 19,000-acre property. By 1961 the original six estates and the recent purchases had been joined by 13 new ones and some 5,516 acres of newly planted tea had more than doubled the total acreage. It then purchased the estates of the Buret Tea Company at Chemosit and Kaptien to add a further 2,035 acres of tea, giving a total tea acreage of 12,134.[11]

Apart from the development of the Timbilil land, new development virtually ceased over the next 15 years and, in 1976, the total acreage stood at 13,500.[12] During this period the parent company merged with Liebig Meat Extract Company to form Brooke Bond Liebig in 1968 (the year after Rutter retired). So far as Kenya was concerned, it was decided to obtain a stock exchange quotation in 1972, when approximately 12 percent of the equity of the Kenya business was sold on the Nairobi Stock Exchange and the head office moved to Nairobi. Brooke Bond already owned some coffee farms near Limuru, had built the Tea Hotel in Kericho and had started a cinchona plantation there. This trend to diversification of the company's activities continued with new ventures in fishing flies, canned tomatoes, agricultural chemicals and carnations, while the move into tea trading in Mombasa was expanded. This created the impression that the mainstay plantation business had lost its priority at head office to tea marketing and new activities. However, the Tea Hotel was sold in 1971 and the fishing fly business was closed down following a fire.

More seriously, Brooke Bond's management of the local tea market through the Pool came under challenge from other producer members, and especially KTDA, with the result that it had to give up its role in 1977 and sell the packing factory to the new government-sponsored organisation the Kenya Tea Packers Association, under the initial management of KTDA. There was a book loss of £1.3 million and the managing director was retired.

In 1984 the Brooke Bond parent company was purchased by Unilever in a contested takeover battle and a new era commenced in Kenya. The diversification of investments was reversed, stage by stage: the cinchona development was stopped in 1984 and progressively returned to tea; canning and agricultural chemicals went in 1985, sisal in 1993, coffee in 1995 and flowers in 1998.The head office returned to Kericho and the focus returned to the growing and manufacture of tea, with the start of a ten-year programme of new planting to add some 3,700 acres on available land around the existing estates. By 2000 the tea acreage in Kericho stood at 18,730. The attraction of Brooke Bond to Unilever was the purchase of a well-known consumer staple, so that there was some surprise when the well-known Brooke Bond brand was displaced by Unilever's Lipton brand in its international marketing. The tea trading unit in Kenya was absorbed into the parent's wider trading operations. Observers in Kenya wondered how long Unilever would regard its investment in actual tea growing as a necessary element of its global marketing business.[13]

Notwithstanding the dispute with the Kipsigis over the Brooke Bond land transaction, the Kenya government resolved to press on with the disposal of the rest of Kimulot to tea company investors. In 1949 it invited bids from 13 firms – most of which had an established presence in India – for 5,500 acres lying between the Maramara and Koruma rivers, as well as for 6,500 acres at Changoi. James Finlay had been interested in the adjoining piece of Crown land since the 1930s and it decided to put in bids for both lots, albeit with a preference for Kimulot. Following an appearance before the

Lands Board, Villiers-Stuart was informed that the tender for Kimulot had been successful at £26 an acre. There were a number of families who had already been living on the land for many years and they were soon joined by others seeking to claim historic possession. Villiers-Stuart took a stand that the government must provide James Finlay with vacant possession, which proved difficult to deliver. As a result, James Finlay did not gain effective access to the land and commence development until 1952, with tea planting commencing in 1954 – the year after Villiers-Stuart's retirement.[14] The company was concerned to re-establish a good relationship with the local community and it arranged to grant lifetime occupancy to those families that had been residing on Kimulot for a long time. Subsequently, a more comprehensive solution was evolved in 1970 when the company agreed to excise some 600 acres from the new Chemamul estate and sold 2.5-acre plots to 205 families at a concessionary price, together with a 75-acre community tea plot. This development at Chepchabus is supervised by James Finlay, who also purchase the harvested green leaf.[15] Later, in 1977, the Kipsigis County Council invited James Finlay to manage and increase the tea acreage of its Kabianga estate (formerly a KTDA nursery).[16]

When the Tea Restriction Scheme came into force in 1933, James Finlay had a much larger area planted to tea (5,000 acres) than Brooke Bond. It was anxious to resume development as soon as possible after the war. However, notwithstanding the release of askaris from war service, there was still a severe labour shortage, as an increase in income in the reserves from war gratuities had led to ex-soldiers being reluctant to join the migrant workforce in the colony. Equally serious was the ban on exports of tea seed by India and Ceylon as a means of restricting tea development in Africa, since the supply of local seed bearers was very limited. Over the six years from 1947 to 1953 James Finlay only succeeded in planting 470 acres of tea; in the next five years, 2,156 acres were planted. By 1961 its tea acreage stood at 9,000, close to double the pre-war figure.[17]

In response to the shortage of tea seed, Hugh Thomas was despatched to Ceylon in 1949 to investigate the new technique of vegetative propagation (VP) that had been developed at the Tea Research Institute of Ceylon and was being tried out on one of the James Finlay estates. On his return, the Kericho estates established trial nurseries, which became the basis for the company's much earlier and faster adoption of selected varieties and VP compared with developments at Brooke Bond. After a pause, there was a renewed expansion drive which brought the James Finlay tea acreage to 12,000 in 1976 and to 14,000 by the end of the century.

With the establishment of a corporate office in Nairobi in 1952, Sir Colin Campbell moved from India to take charge of James Finlay's Kenya interests, leaving a superintendent in Kericho to oversee the tea estates. Unlike many in the industry, he played an active part in the affairs of the tea industry and took a notably cooperative attitude to the newly emerging smallholder sector, for many years sitting on the board of KTDA. James Finlay invested in some of the first KTDA factories. There was a marked effort to meet Kipsigis' aspirations; in addition to the Chepchabus settlement and the management of the Kabianga estate, already referred to, James Finlay subsequently took on the management of the failed Nyayo project at Sinandet, which was transferred to the Kipsigis Multi-Purpose Cooperative at President Moi's behest, and also the management of the Mau Forest estate after Moi forced his protégé Mark Too to divest to the Kipsigis cooperative.

The Mau Mau emergency was brought to a formal end in 1960 and the British government decided to accelerate the colony's preparation for political independence. A constitutional conference in London that year led to the unexpected decision to grant independence in 1963. This timetable presented a big challenge to the tea estate sector, with its implication that there would need to be a rapid replacement of expatriate European managers by Kenyans. James Finlay and Brooke Bond had contrasting approaches to the new situation. The precedents from South Asia were not encouraging: in India there had been onerous employment restrictions and widespread takeover of tea companies by Indian business interests; in Sri Lanka there had also been outright nationalisation. The latter was not a threat in Kenya, but there was a serious challenge to the management of the estates for which the industry was ill prepared. The efficient replacement of the colonial service in Kenya over the independence period was also a significant challenge. More generally, there was the question of how the leadership of the industry would relate to the new political dispensation. Colin Campbell of James Finlay refused to agree to numerical targets for the reallocation of management jobs to Kenyans, meanwhile vigorously recruiting and promoting Kenyans for positions within the firm, and relying on his connections in the presidential office to argue his case. The situation at Brooke Bond was different. The company was dominated by its autocratic deputy chairman, TD Rutter. A succession of weak superintendents in Kericho left the field clear for him to dominate company activities from London, which was not conducive to adjusting to the new political climate in Kenya. Brooke Bond was notably uncooperative over assisting in the early development of smallholder tea and declined to invest in or manage any of the first group of KTDA factories, although it participated later on. Brooke Bond was unable to make an argument against the demand for numerical management targets and agreed a timetable to reduce its expatriate managers from 74 to four. James Finlay reduced its number of expatriate managers from 70 to only five, achieving a similar outcome, but several memoirs attest that morale during the transition was much higher at James Finlay, as it was felt that the company was managing the process with more consideration for the individuals involved.

Campbell initiated a number of business acquisitions in Kenya for James Finlay that were perhaps modelled on the traditional agency houses in India, investing in clearing and forwarding, import agencies, cotton trading and retailing. This expansion continued on a larger scale when he assumed chairmanship of the group in Glasgow in the 1960s, fuelled by the sale of the India tea plantations to the Tata group. James Finlay then came under predatory attack, with a near 30 percent hostile shareholding by Slater Walker, but the threat collapsed in the secondary banking crisis of 1973. The outcome was that the stake was purchased by the privately owned Swire group. The non-tea interests were progressively disposed of and the Kenyan headquarters returned to Kericho in 1991. Swire went on to acquire full ownership of James Finlay in 2000. It is a firm with extensive experience of managing overseas businesses in Asia, with a widely admired management team that has been in a position to take a long-term view of its investments.

Both James Finlay and Brooke Bond originated as family-controlled businesses, with the former drawing on a wider circle of families with a tradition of overseas employment and Brooke Bond relying on a narrower social circle. James Finlay was quicker to see the potential of vegetative propagation of outstanding tea varieties, and also of instant tea, and appears to have been better managed down to the 1980s. Brooke Bond led on labour conditions and welfare, but both companies were challenged by the changed working environment of political independence. Both of them fell for the fashion for business diversification in the 1970s, but only James Finlay has held on to its successful flower export development. The subsequent acquisition of James Finlay by the Swire group and of Brooke Bond by Unilever has refocused the two Kericho companies on their core tea business.

The third piece of Crown land at Kericho to be put up for sale by the government was at Changoi. It attracted the attention of an Indian agency business that had first been tempted by prospects in Kenya in the 1930s and was now seeking to establish a more substantial presence. The George Williamson group of companies had a typical foundation story: two Williamson brothers, one of whom was a tea estate manager, the other a Ganges riverboat captain, along with a third man named Magor, the manager of a hotel in Calcutta, formed a partnership in 1869 to manage tea estates, sell tea and invest in tea properties. This dual activity of managers and investors was a distinctive feature of the Indian managing agency system; in this case, the enterprise remained as a partnership in Calcutta until 1954 and in London until 1974, when they were incorporated as companies. In London, one family, the Magors, from a low point of 5 percent in 1949, came to own the whole of the business over the following 30 years and now trades under the name of Williamson Tea.[18]

In 1934, RK Magor and his son Richard paid a visit to Kenya and were impressed by the development being undertaken by Brooke Bond and James Finlay; however, they concluded that it would be more costly to bring tea to the world market from Kenya than Assam and advised their partners against investment. After the war, uncertainties over the future of the industry in India prompted a further visit from one of the London partners, EJ Nicholls, in 1947. He decided to set up an office in Nairobi and to seek managing agency mandates from Indian tea companies that were looking to invest in East Africa. A number of agencies were secured in Tanganyika and Richard Magor moved to Nairobi to take charge of their interests.[19]

The first Kenyan agency was Kapchorua in Nandi, where a number of investors connected with Williamsons purchased the property and floated it as the Kavirondo Tea Company in 1948. However, this was a high point because the London partners became increasingly sceptical of the merits of investing in East Africa. Their initial fears over the security of their interests in India had proved unfounded and the Indian estates companies were proving unwilling to invest in Africa in preference to opportunities near at hand. The East African business was not yet viable and Magor had a struggle to prevent the outright closure of the Nairobi office, let alone to build up a worthwhile portfolio. Kapchorua was starved of development capital and only 490 acres of tea had been planted by 1959. The corporate interest had remained at 27 percent, but was augmented to control by family investment and the estate had been expanded to 1,700 acres by 2000. Kaimosi Tea Estates was incorporated in 1947 as a venture with one of the original Kericho settlers, Floyer, who had some 80 acres of tea. It purchased six farms from the government amounting to 2,000 acres that had originally been part of the 1919 Soldier Settlement Scheme, let on 999-year leases, but which had failed and the land had reverted to the government. The project was initially precarious, due to shortage of capital, and had only 410 acres of tea by 1959. Williamson gradually acquired full control and the tea acreage was increased to 760 in 1976 and to 1,100 by the end of the century. Tinderet estate was started by James Warren from India under the management of Williamson and there were 134 acres of tea by 1959. James Warren developed the estate to 1,000 acres by 1986 and then sold to the local Indian-owned Sasini group. It was then sold to two Kenyan investors, Arap Leting and Mark Too, under Sasini management until Williamson then purchased a 76 percent controlling shareholding in 1993 and increased the acreage to 1,500 by the end of the century.[20]

Development in Kericho was also uphill work for Williamson. The agency for Kaisugu estate was arranged with Colonel Brayne in 1949, after two years' solicitation, but he died within days of its completion. His widow, now chair of the company, was a notoriously difficult personality and she sabotaged both equipment and records before leaving. An uncomfortable relationship with the family continued for a while until the agency was terminated. This was a distraction from a much more significant event. Magor had become aware in 1949 of the government's plans to auction Crown land at Changoi near Kericho; however, the Williamson partners were very sceptical about the prospects and the proposed conditions. Their feeling was that overseas investment interest in Kenya was not strong; that development costs would be high; that official controls over domestic tea prices to keep them below world prices was a disincentive; that labour supply was uncertain; that freehold was not being offered; and that the possibility of rent reviews was a further discouragement. They therefore refused to participate in an auction, but suggested that a bilateral negotiation would be considered. This robust line paid off, since the government was determined to secure a major new investor in preference to just allowing expansion by the existing incumbents in Kericho. Richard Magor was able to negotiate the purchase of 4,000 acres in 1951 at £7 an acre, payable over ten years, notwithstanding the fact that James Finlay had bid £26 an acre. Nevertheless, the episode is a reminder that there were real perceived risks in undertaking new agricultural development at this time.[21]

There was an extraordinary coda to the Changoi story in that, when the government came to survey the land, it was discovered that the area that had been sold was only 3,000, rather than 4,000 acres. Williamsons took a tough line over the shortfall and, after protracted negotiations, the government was eventually able to purchase 500 adjacent acres from James Finlay at £26 an acre and pass this on to Williamsons at the agreed price of £7 an acre. The company was also able to purchase a small neighbouring farm planted with 26 acres of tea, planted by Harry Borman in the 1940s The combined properties were initially developed to carry 527 acres of tea by 1959 and expanded to 1,353 acres by 1976 and to 2,200 acres by the end of the century. At its peak, Williamson had a management role over 16 tea estates in Kenya and Tanzania, but the business was steadily rationalised until, by 2000, it had been consolidated into the ownership and management of the two estates in Kericho and three in Nandi, with a combined tea acreage of 6,500. [22]

Two small Kericho estates were sold to prominent Kenyans in the Moi era. In 1986 President Moi himself purchased Kaisugu Estate for £2.4 million and installed a Brooke Bond manager, John Popham, to manage the property. The year before the neighbouring Mau Forest estate was purchased from Gregor Grant by Mark Too, chairman of Lonrho Kenya and a close associate of Moi's. It was then placed under Popham's management. In 2000 the estate was sold to a Kipsigis cooperative at the instigation of the President and then managed on its behalf by Finlay.[23]

In 1938 eleven Sotik farmers had been allocated 'starter' licences under the International Tea Agreement for 330 acres of tea, and by 1943 six of them had planted some 309 acres and they were awarded a further 530 acres in 1945. From these modest beginnings a number of small estates were brought together in 1948 as the Sotik Tea Company: Arroket from Jan George, Kaptembere from Commander Letts, Monieri from Maitland Edye. In the following year a controlling interest in Sotik Tea was acquired by the Wemyss Development Company, which had been set up by the Wemyss family following the nationalisation of their coal mines in Fyfe, in order to invest the compensation proceeds.[24] Brian Shaw developed an estate at Kipkebe and African Highlands began to develop Sotik Highlands. By 1952 the Sotik district had some 1,000 acres of tea planted, expanding to 2000 acres in 1956. An ownership breakdown is available in 1959,[25] showing:

Sotik Tea Company	926 acres
Sotik Highlands	394
Kipkebe	454
9 others	634
Total	**2,408**

In 1964 the Sasini group purchased a stake in Kipkebe and then full ownership in the following year. Subsequently it purchased Keritor with 206 acres and by the end of the century the combined tea acreage was 3,345. In 1984 the Sotik Tea Company purchased Sotik Highlands from James Finlay and by the end of the century the combined property had 4,455 acres of tea.[26] Thus there were two well-established estate enterprises in Sotik by 2,000 with some 7,800 acres of tea adjoining the burgeoning smallholder sector in neighbouring Kisii District.

Apart from the Crown land excisions in the Kericho area, the only significant area of land that was suitable for tea development comprised the farms in Nandi that had lobbied successfully for planting licences during the period of the Tea Restriction Scheme. This district became the main focus of interest for sterling tea companies from India and Ceylon looking to invest in Africa after the war. The efforts of George Williamson to establish themselves have already been noted. Another early development involved the small Mokong estate of EG Myers, which was sold in 1947 to become the Nandi Tea Company, owned first by the Uganda Company and then by the Mitchell Cotts group. It had reached 1,300 acres of tea by 1959, 1,700 in 1976 and 2,380 at the end of the century. However, the principal investor in Nandi has been the group now known as Linton Park. It is formed by the combination of several historic strands of Indian and Sri Lankan managing agency and tea estate businesses. The Kenyan investments form part of a much larger group which owns tea estates in Malawi, India and Bangladesh, aggregating some 38,000 ha of tea, as well as agricultural and industrial investments that have been brought together by a remarkable Canadian businessman, Gordon Fox, and ultimately controlled by his private foundation. The originating businesses comprise two estate agency businesses from Calcutta and a Ceylon estate company.[27]

Alex Lawrie was part of the wave of entrepreneurial Scots who set up some fifteen substantial managing agency businesses in India between 1850 and 1890, during the development phase of the Assam tea industry. He went to Calcutta in 1862 at the height of 'tea mania' and set up a managing agency partnership, Balmer Lawrie, in 1867. A London partnership, Alex Lawrie, was established in 1878. In both cities the partnership was maintained until the 1920s; subsequently, they became quoted public companies – Alex Lawrie in 1935 and Balmer Lawrie in 1936. The post-war loss of confidence in the prospects for foreign business in India led to Balmer Lawrie losing a number of important managing agency accounts, as estates came under the control of Indian investors. In 1956 control of Balmer Lawrie itself fell to Steel Bros. (the Goenka Family), and the company was subsequently nationalised in 1972.

Gordon Fox was first involved with Alex Lawrie through his shareholding in a small tea company, The Septinjure Bheel Tea Co, with a single estate in Cachar, north-east India, for whom he was managing agent. In 1961 a dissident shareholder tried to have Alex Lawrie dismissed and, when he failed, Fox purchased his stake. In 1964 the estate was sold and the company became a shell and its name was changed to Camellia Investments. Fox joined the board and increased his stake to 30 percent by 1967 and to 51 percent by 1969, when he became its chairman. Meanwhile, he became an 11 percent shareholder in Alex Lawrie, arising out of a joint venture. This was the era of asset-stripping investors on the London Stock Exchange and one such, Oliver Jessell, began stake-building in Alex Lawrie. In order to escape the embrace, a block of shares representing 31 percent was offered to Walter Duncan Goodricke in 1967, which enabled it to mount a bid for the rest of the capital (including Fox's stake), and Jessel departed the scene. Fox was already an investor in Walter Duncan Goodricke and he became a director in 1976.

The Duncan story also began in Calcutta in 1859, initially in cotton piece goods, tea agencies and jute. Duncan Brothers in Calcutta became a major tea agency, but it fell under Indian control in the 1950s through the Goenka family. Meanwhile, back in Britain, the Walter Duncan partnership had merged with another agency house, CA Goodricke, in 1949 to form Walter Duncan & Goodricke, which was floated on the London Stock Exchange in 1951. In 1977, eight of the sterling tea companies (comprising 17 estates) managed by Duncan Brothers in India decided to break away and join the group of companies controlled by Walter Duncan & Goodricke. This entity comprised 35 estates in India with 40,000 acres of tea.

Following the purchase of Alex Lawrie by Walter Duncan & Goodricke in 1967, a strategy was followed over the next ten years of increasing the cross-shareholdings between the individual tea companies (which had the resources to do so), in order to make the group as a whole bid-proof. From under 30 percent the holdings increased to 50 percent, with Camellia and Walter Duncan & Goodricke as active players as well. As a result of the 1965 war between India and Pakistan, Indian rupee companies were sequestered. However, the sterling companies in what was to become Bangladesh managed by Duncan Brothers and Octavious Steel (also Goenka) then gravitated to Walter Duncan & Goodricke and were amalgamated as Longbourne Holdings. Following the creation of Bangladesh in 1971 there was no prospect of the Goenkas getting back the rupee companies, with the result that they were also sold to Longbourne. The outcome of these developments was the acquisition of 14 estates in Bangladesh, with 20,000 acres of tea.

The third strand in this complex history had its origins in Ceylon and was the pathway to Kenya for the Linton Park group. Eastern Produce Estates was formed in 1887 to purchase the assets of a failed coffee company and to plant tea. In 1943, and again in 1952, Eastern Produce greatly expanded its tea interests by purchasing two managing agency companies in Colombo; in 1955, it also bought the principal managing agency in Malawi. Over the following years, Eastern Produce built up ownership of nearly 14,000 acres of mature tea in Malawi. Development in Kenya, in the Nandi district, proceeded more slowly. In 1948 it began to establish an agency business, but also purchased Chimomi farm, which was subsequently planted to 1,600 acres of tea. The initial managed estates were 1,000 acres at Kapsombweia, on behalf of Empire of India, Singlo and Dooars, and 890 acres at Kibabet, on behalf of Standard Tea, Bogawantalawa and Pundaloya. Other agencies acquired in the 1950s were for the 680 acres of Kip Koimet, for Craighead Investments, but where Eastern Produce was also an investor. In 1972, Eastern Produce acquired a stake in Kakuzi Limited; this old established business had originally been a major sisal grower founded by well-known Kenyan pioneers before the First World War.[28] After the Second World War it purchased Kaboswa farm in Nandi and planted tea, and then purchased the neighbouring Siret tea estate, which already had a factory, leading to an estate of 1,600 acres of tea. Eastern Produce also bought out another early development, East African Coffee Plantations, where GR Mayers, who had come to Kenya from Australia initially to grow sugar, had eventually succeeded in establishing two tea estates at Kepchomo and Savani with 600 acres by 1959 and 1,800 acres by 1976.

Jessel Securities had also built up a hostile stake in Eastern Produce between 1969 and 1973, but on its collapse in 1974 the 30 percent stake was purchased by Walter Duncan & Goodricke in 1976 after all the Sri Lankan assets had been nationalised in the previous year. By stages, Eastern Produce became a wholly owned subsidiary. There followed a regrouping of the tea companies in 1978 in order to establish a holding company, Lawrie Plantation Holdings, controlling management companies for India, Bangladesh and the new African interests. Initially, Camellia owned 46 percent of Lawrie. In 1979 the separately quoted companies were all delisted and in 1990, under a scheme of arrangement, Camellia held 71 percent of Lawrie, which in turned owned the whole capital of subsidiary companies. By 1999 Camellia owned the whole share capital of Lawrie and was itself a quoted public company with some 40 percent of its capital publicly owned.

Eastern Produce changed its name to Linton Park in 1990, the name of the Palladian mansion in Kent which houses the corporate headquarters of the group. In 1976, after numerous acquisitions, the Nandi estates comprised 8,400 acres of tea; by 2000, its area had been increased to 12,000 acres. In addition, there were long-term management agreements embracing some 4,000 acres of tea owned by private investors and arrangements to purchase green leaf from 3,500 acres of smallholder farms.[29]

Towards the end of the war there were some 850 acres of planted tea in Nandi District. As the new investors began to develop their properties the tea acreage expanded to 2,582 by 1952. The further initiatives by the estates sector reviewed above then took the planted tea acreage to almost 21,000 acres by the end of the century.[30]

The Linton Park management agreements with privately owned tea estates point to a special feature of tea ownership in Nandi. The sale of settler farms to the new Kenyan elite has resulted in the emergence of an influential group of investors who have placed the management of this specialised crop in the experienced hands of major tea companies, to their mutual benefit. Thus the 800-acre Koisagat estate passed into the hands of Charles Karanja of KTDA and the auditor general, Njenzi; the 660-acre Kibware estate was bought by the Kericho MP Henry Kosgei, who also owns the 317-acre Kaprochege estate and Kapkubor; the 257-acre Kaptino estate is owned by the Sarah Boit family; the 406-acre Chematin estate was bought by Thomas Ngana; and the list continues. The tea companies have also been ready to purchase green leaf from KTDA smallholders in the area, relieving pressure on its own factories as well as providing price competition. In this way, Linton Park, Williamson and Nandi Tea Estates have been able to establish a productive relationship with a circle of smaller private tea estates and with the much larger smallholder community.

Limuru did not provide the same opportunities for investment by the sterling tea companies as Nandi, but there has been a repetition of the pattern of settler farms passing into ownership of the Kenyan elite. In the early post-war period Brooke Bond consolidated its leading position in the district with its Mabroukie factory purchasing green leaf from a growing number of farmers with tea plots. Its main supplier was the Limuru Tea Company, which had a quotation on the Nairobi Stock Exchange and in which Brooke Bond built a 52 percent controlling stake. By 1959 there were 24 farms with licences to plant tea and 19 of them had planted 1,262 acres between them, as against the 578 acres of tea on Mabroukie itself. After political independence the new Kenyan elite began to purchase farms in Limuru and there are now more than 50 farms growing tea. One consequence is that the original grievances over displacement by settler farmers have been almost entirely erased by the arrival of new, mostly Kikuyu owners who have been robust in dealing with historic claimants. Three new tea factory companies have been established: Maramba also has a nucleus estate of 250 acres of tea and its investors are led by Paul Thiongo, a former director of agriculture; Ngorongo is led by Njange Karume, a transport businessman and MP; and Karurana is led by Philip Ndegwa, a former Treasury permanent secretary. The new tea farmers included former Attorney General Karuga, with 300 acres; Joseph Koinange, with 40 acres; Kanja Promote, a transport businessman, with 80 acres; and Ngethi Njoroge, ex-high commissioner in London, with 40 acres. The tea farmers now had the choice of where to sell their green leaf harvest, and Brooke Bond has not had an altogether smooth ride in trying to reconcile its increasingly strict production standards with independent-minded growers. The case histories of Francis Wacima and Bill Brown are illustrative.[31]

Wacima originally worked for the railways in security and, in 1968, decided to leave Nairobi for the benefit of his children's education and purchased 20 acres in Limuru from solicitor Wollen, with bank loans, and with a green leaf agreement with Mabroukie. Around the time of the millennium a number of the outgrowers became dissatisfied with the payment arrangements and formed the Limuru Tea Farmers Association to negotiate with Brooke Bond, which proved difficult. The upshot was that about 40 farmers, including Wacima, broke away, switching first to the KTDA factory at Kiambaa, and then to Maramba.

Brown has 43 acres of tea and sells his green leaf to Mabroukie. His father was manager of the Limuru Tea Company and he himself worked there for 30 years. When he sought to purchase the plot owned by Moss, the dentist's widow, in 1988, local Kikuyu tried to disrupt the deal. At the time, Brown was managing the coffee estate of J Kibe, who intervened and stopped the harassment. Brown has put in a dam and irrigates his tea to obtain a higher yield, and has built a village on the farm with 35 houses, as well as employing casual labour.

By 2000 there were nearly 6,000 acres of tea at Limuru owned by members of the KTGA, and there would also have been several hundred acres owned by non-members, before taking into account KTDA smallholders in neighbouring Kiambu and Maranga districts. Limuru has the distinction of being where the first tea was planted in Kenya – by a pioneering settler, and where a tea company from India made the first investment by purchasing the Caine farm and developing Mabroukie estate. The subsequent arrangements whereby Brooke Bond purchased green leaf from surrounding small estates to process in its factory established a prescient model for the future when, after independence, members of the Kenya elite invested in tea farms. However, the presence of the Mabroukie estate sets a benchmark for tea cultivation standards in the district and it also challenges Unilever to sustain a working relationship with the outgrower community.

Released from the restrictions of the ITA at the end of the war, the established tea companies and new entrants invested heavily and the sector expanded from 6,500 to 35,000 ha by the end of the century. This growth has continued and the area under estate-managed tea increased further to 90,000 ha over the following 15 years, by dint of bringing into cultivation all available land owned by the estates that was suitable for growing tea. It should be noted that this total figure also includes the larger tea properties owned by the Kenya elite and managed by estate companies in Limuru, Nandi and Kericho.

NOTES TO CHAPTER 9

[1] Cf. chapter 1, footnote 19.

[2] MP/BB/1/6, Chief Secretary to Provincial Commissioner Nyanza, 15 August 1947.

[3] MP/BB/1/6, Chief Secretary to Provincial Commissioner Nyanza, 10 December 1946.

[4] MP/BB/1/6, District Commissioner Kericho to Provincial Commissioner Nyanza, 3 September 1947.

[5] MP/BB/1/6, Local Native Council minute, 11 January 1949.

[6] MP/BB/1/6, District Commissioner Kericho to Provincial Commissioner Nyanza, 25 January 1949.

[7] MP/BB/1/6, Member for African Affairs to Provincial Commissioner Nyanza, 7 April 1949

[8] MP/BB/1/6, BA Ohanga to Provincial Commissioner Nyanza, 30 October 1950.

[9] MP/BB/1/7, Timbilil. This account is a summary of government files on the matter.

[10] MP/ITA/1/1, Memorandum by Director of Agriculture, June 1938. This sets out the area of each estate and the new planting allocations for the first period and for the renewal.

[11] MP/Library/8, *Tea Estates in Africa*, published by the tea brokers Wilson, Smithett in 1961, which lists the acreage of each estate.

[12] MP/KTGA/1/2, statistical report for 1976.

[13] MP/1M/3/1, Davies interview, 12 August 2003, and Laurent interview, 20 August 2003. Ironically, it is the slow growth prospects for tea consumption in its main markets that has prompted Unilever in 2020 to signal that it is considering the disposal of all its tea businesses: *Financial Times*, January 2020.

[14] MP/Library/3, Thomas gives a summary account of the Kimulot issues, p.55. See also MP/1M/5, Research Diary, Paterson. The purchase price is deducible from the Changoi transaction.

[15] MP/Library/5, Soy interview, 12 August 2003.

[16] MP/AH/4, James Finlay house magazine, 1976 and 1978.

[17] MP/Library/3, Thomas, Appendix 5. MP/Library/8, Wilson, Smithett, for 1961.

[18] *Williamson Magor Stuck to Tea*, Peter Pugh, Cambridge Business Publishing, 1991.

[19] The Rajmai Tea Company purchased Chivanjee in the Southern Highlands, and subsequently two neighbouring estates, Kiganga and then Musekera, from Brooke Bond, who had been managing the former German property during the war. In the Usambara Mountains the managing agency of Ambangulu and Dindira was obtained.

[20] MP/EDK/1/6.

[21] The Kipsigis Changwony clan originally lived in this area, which became referred to as Changoi. Its leader in the 1950s, Arap Changwony, became a supervisor on the tea estate. MP/EDK/1/6.

[22] MP/1M/2/1, Harrison interview, 22 May 2014.

[23] MP/1M/4/5 Interview with John Popham, 2003.

[24] MP/Library/6, Karmali, Joan. *The Story of the Sotik Tea Company*, Nairobi, 1991.

[25] MP/KTGA/1/3, Tea Board of Kenya, Tea Estate Acreages, 1959.

[26] The transaction was much regretted by the Finlay management and it was said that Sir Colin Campbell wished to establish a market value for Kenya tea estates as a political insurance. MP/1M/5, Tea Research Diary, Davies and Course. The prominent Moi minister, Nyachai, was and remained a shareholder. The sales price was £2.18 million plus a special dividend of £1.1 million, all equivalent to a price of £8,200 an acre, which was twice the value at which James Finlay had just revalued its Kericho estates. MP/AH//2, Annual Accounts of African Highlands Produce Co. Ltd.

[27] MP/Library/4, *Camellia – The Lawrie Inheritance*, Michael Manton, published privately, 2000.

[28] Donald Seth-Smith, one of the founders and chairmen of the Muthaiga Country Club, Lord Cranworth, author of *Kenya Chronicles*, and Mervyn Ridley, his brother-in-law.

[29] Tea Board and KTGA schedules show that the 1959 area of 1,524 acres had been developed by acquisition and new planting to 8,360 acres in 1976 and to 12,079 in 2000.

[30] MP/KTGA/2/1 for 1943 acreage. MP/Sales/2/4 for 1952 acreage. MP/KTGA/1/2 for 2000 acreage.

[31] MP/1M/5, Tea Research Diary, interviews with Brown on 23 August 2003, Wacima and Wanjui Cliffe on 28 August.

Clockwise from above: Kitumbe Estate, Kericho – renewed growth following pruning; Kimugu Factory, Kericho; Stephen Musongo, manager Kaproret Estate, Kericho

Above: Tea plucking on smallholder farm, Nyeri

Below: Smallholder farms, Nyeri

Above: Smallholder farms, Nyeri

Below: Seedbearer at Limuru, planted in 1903

Tea leaves, flowers and seed from Siret Estate, Nandi.
Painting by Margaret Booth-Smith

10

ENTER THE SMALLHOLDER

Searching for a Policy

African farmers made small plantings of tea in the 1930s in both Kenya and Uganda, using seedlings taken from the new estates, and, in due course, making sun-dried tea for local consumption. These initiatives were first noticed officially in Uganda in 1936 because they constituted a potential infringement of the legislation passed by the East African governments to licence and control tea growing under the International Tea Agreement, not to mention the Excise Ordinance. There was also a more political dimension, as forecast by the Governor, Sir Philip Mitchell: "When we discussed the matter in 1937, at the Colonial Office, I told you how awkward things would be as soon as a Muganda landlord demanded to be allowed to grow tea on freehold land in Buganda for Baganda to drink. That difficulty has not yet arisen but I am certain it will arise one of these days."[1] And indeed, in 1946, returning askaris, who had served with the King's African Rifles in the Burma campaign and seen tea in Ceylon, had noted the opportunity: "Sir, as you are our adviser, I should say that the time has now come to help in giving us permission to grow tea in the same way as we are permitted to grow coffee, sugar cane and other products."[2]

The issue of whether tea should be grown by African farmers was raised at the 1945 Conference of East Africa Governors, which prompted Kenya's Department of Agriculture to undertake a survey, when it was discovered that small quantities of tea had been grown in Fort Hall for many years: in Chief Joshua's location since 1931 and in Chief Njiri's since 1933, followed by others since 1939. In all, there was the equivalent of some 13 acres of tea.[3] Although the provincial staff in Central Province and Nyanza were initially sceptical about the prospects for tea growing, a more positive attitude was taken by the Director of Agriculture. This was crystallised in an important memorandum by him in late 1947, '*The Future of the Tea Industry*'.[4] The general policy approach was clearly articulated:

"It had been widely held in the past that the African was not capable of producing plantation crops on the same scale and of the same quality as Europeans, but this theory can no longer be held. African produced arabica coffee in this country more than holds its own in point of quality in comparison with European coffee. Similarly pyrethrum of good quality and a high rate of yield is being produced, and it cannot be doubted that the same would apply to tea, provided that it is grown under the same general conditions, that is to say, that the areas in which it is planted by natives are proved to be suitable for tea production and that the growing of tea is sufficiently closely supervised by competent officers with adequate powers to ensure that cultivation and all other aspects of production are properly carried out."

The Director went on to envisage the estate sector expanding from 14,000 acres to 60,000 to 70 000 acres, while, for smallholder tea, "the development of several thousand acres might be possible eventually in the Native Reserves surrounding Mount Kenya and those adjoining Kericho and Sotik districts."

The next step was to consult the estates sector through the KTGA, since the Department of Agriculture was very conscious that it lacked expertise on tea, and also there was the looming issue of the financing of factories for African smallholders.[5] Although the formal KTGA minute on the paper recorded its agreement with the general principle of tea being grown by African farmers, there was no response to the invitation to erect a factory to process such tea, and there were lively apprehensions over such matters as disease control, leaf stealing, jeopardised labour supply and relationships with outgrowers. Nevertheless, it agreed to collaborate in the preparation of enforceable cultivation rules.[6] Three main factors influenced this lack of enthusiasm for the aspirations of the Department of Agriculture. First, the tea companies were aware that smallholder tea growing had not been successful in Ceylon, and this outcome was transposed to Kenya, without reflecting on its causes. Second, Brooke Bond's experience of purchasing green leaf from European outgrowers in Limuru had been stressful and did not provide an encouraging precedent. Its original justification had been the difficulty of acquiring enough land in Limuru and it also helped to save on development costs at the pioneering stage, so that it represented a speedy route to getting Kenya-grown tea into the local consumer market; later, the restrictions on development in Kericho under the ITA provided an incentive to make the best of the relationship with the far from docile outgrowers. The conclusion was that an integrated factory estate was by far the preferred mode of operation when the choice was available. The third influencing factor was the feared impact of political agitation on an institutional structure comprising a company factory purchasing leaf from African farmers. Here, the recent experience of a wartime initiative to establish a dried vegetable factory at Karatina to process locally grown vegetables for military supplies seemed to provide a warning. Although it was a successful operation, an agitation developed post-war because the factory had been built on reserve land (as opposed to Crown land) and the outcome was that the factory was demolished and the land returned to its original owners.[7] To the fearful, the Karatina episode raised questions over the security of leaf supply to an investor-financed factory.

The outcome of the consultation was to reinforce a line of thinking that was already present within the Agricultural Department, namely that tea growing by African farmers should be fostered under the sole aegis of the Department. A pilot area was selected in the Mount Kenya foothills at Kangochi in 1949, well away from the existing tea areas, with funding to be provided under the African Land Utilisation Scheme. The Central Province agricultural officer, Grimy Gamble (as he was popularly known), became the tea project champion and he was sent on an investigative mission to India and Ceylon in early 1950 to gather all available information on best practice in these countries, and to formulate guidelines for Kenya. His report was published for the benefit of the industry as a whole.[8] Gamble discovered a confusing variety of practice, strongly influenced by company tradition, and as yet only partially moderated by scientific research findings. Nevertheless, he saw enough of the new technique of vegetative propagation to form a firm view that this was the preferred route for Kenya, especially as reliable seed supplies were limited. Most importantly, he also reached clear views on how to avoid the problems of the Ceylon smallholder sector: these were essentially that growers should be closely supervised by the Department of Agriculture under formal powers to insist on good husbandry and leaf inspection standards. He was firm that African tea must be of export quality and that growers should receive regular payment for green leaf, with an end-of-year balancing payment. Following Gamble's return a departmental paper illustrated how far its ambitions had escalated since the cautious vision of 1945. For Central Province alone it estimated that the ecological area suitable for tea growing was some 246,000 acres, and that under a mixed farming regime, with half an acre of tea per family as a principal cash crop, alongside the sale of milk, vegetables and wattle, approximately 10 percent of the area, or 25,000 acres, might be devoted to tea. It was envisaged that tea factories would be financed with Colonial Development & Welfare funds.[9] The first tea seedlings were distributed from the Kangochi nursery in 1952 to progressive farmers, many of whom had already consolidated their fragmented holdings and obtained title, and some 35 acres were planted.

The project then suffered a severe setback when the deteriorating security situation led the government to declare a state of emergency in October that year in order to suppress the Mau Mau rebellion by the Kikuyu in Central Province. The early headlines focused on savage attacks on European farmers, but the rebellion began to assume the characteristics of a civil war in Kikuyu society, between those loyal to the government versus the Mau Mau, which in turn revealed divisions between those with farm land and those without – the latter of whom were readily recruited to the Mau Mau cause.[10] One consequence for the nascent tea programme was to restrict the allocation criteria for planting tea to members of the loyalist Home Guard. However, these men did not have much time to spare for new farming practices; it was also the case that the possession of tea marked out a man as a loyalist and some had their tea bushes slashed and a nurseryman was murdered. A decision was taken "to help such people by using prison labour as a reward for their loyalty" and Mau Mau prisoners were formed into relief gangs to prepare tea plots and to work in the tea nurseries.[11]

The board that had been set up to supervise the project, containing Gamble and Sam Ball, who had retired from Brooke Bond, was nearly overcome by pessimism over the new problems: the slow rate of planting; the difficulty of justifying investment in a tea factory with only some 50 acres planted; and the countervailing concern of letting down farmers who had taken the risk of planting tea. The situation was not helped by a gloomy paper by the director of the Tea Research Institute, Dr Tom Eden, who had previously worked on the smallholder sector in Ceylon: "I came to the conclusion that tea is not a crop that is well adapted to smallholders because of problems of harvesting standards and timely factory delivery."[12] It was decided in 1954 to put the Nyeri scheme into cold storage until the political situation improved. This accounts for the cautious tone of the sections dealing with tea in the otherwise revolutionary Swynnerton Plan that was published in the early part of 1954. *The Plan To Intensify the Development of African Agriculture in Kenya* [13] was in part a creative response to conditions that had given rise to the Mau Mau rebellion, and the

British government provided nearly three-quarters of the initial funding. On tea, it was thought there was the potential for 70,000 acres of smallholder tea within mixed farming regimes, but the plan only allowed for 1,200 acres being planted over the 15 years to 1970. After acknowledging the difficulties, a marker was laid down in the report– "every effort must be devoted to developing sound and practical peasant tea-growing schemes in suitable areas" – and three possible lines of future progress were identified:

- Establishing smallholder schemes adjacent to existing tea factories in Limuru, Kericho, Sotik and Nandi districts
- Establishing a number of cooperatively owned plantations supporting a central factory
- Elaboration of the Nyeri smallholder concept to other districts.

An approach was made to the Tea Board in June 1954 by the Director of Agriculture, Graham Roddan, to sound out the industry over collaborating with smallholder tea development in Kericho itself. The Kimulot Settlement Scheme comprised a total of 240 farms of 20 acres each and had been laid out with roads and water supplies and it was envisaged that each farm might have an acre of tea. The question was whether one of the surrounding tea companies would agree to purchase the green leaf harvest. In approaching the tea companies, Roddan drew attention to the desir-ability on political grounds "that the initiative in this should come from Government and that we should not wait until political pressure and clamour is such that we have to give way," and also that it would be "useful" to have anticipated any recommendations the East Africa Royal Commission might make with regard to tea growing by Africans. Yet he also revealed his own scepticism: "I am far from convinced that tea is a suitable crop for the African."[14] Three tea companies abutted the Kimulot Settlement Scheme, African Highlands, Brooke Bond and Buret, but the formal reply through the KTGA stated flatly "The purchase of leaf is not a practical business proposition" and went on to list a number of other reasons for opposing the proposition: existing factories were working to their planned capacity; leaf transport would be difficult to arrange; labour absenteeism would be encouraged; and the anticipated theft of leaf from the estates. Nevertheless "The Tea Industry does not oppose the growing of tea by Africans, but regrets it is not in a position to assist in a pilot scheme in the Kericho area."[15]

Roddan voiced his disappointment at the negative response by the industry; more importantly he announced that the government would continue on its own. Initially, this was back in Central Province and well away from the established areas, as the tea companies desired, but the Kimulot settlement went ahead with tea in due course.

A technical dispute then surfaced within the agricultural department on the respective merits of establishing plots of tea within mixed farming regimes (a defining philosophy of the Swynnerton Plan) and the alternative of establishing large blocks of tea, albeit owned and cultivated by individual farming families. With land tenure in a state of flux during the emergency, and with the provincial administration exercising strong powers of direction over everyday life – including the use of detainee and communal labour – the tea-block school of development gained a temporary dominance. By 1955 there were 63 acres in Mathira, 86 in Othaya and 16 in Embu, perhaps involving over 300 families on the half-acre standard.[16] Gamble was against the idea of block planting and his views prevailed as the tide of land consolidation swept through the province. Commenting on the Kimulot settlement he wrote: "On no account should the settlers cooperate to plant tea in one block of say 20 to 50 acres. This means that the settler will not have his own tea land, he will not be bothered to visit the communal block daily and undoubtedly the quality of the tea, the cultural methods and cash value will go down. No block planting of any crops (coffee at Embu, cotton in Nyanza) have ever been successful. Each individual settler should therefore plant his tea on his own land adjacent to his homestead area."[17] Much of the block-planted tea was subsequently uprooted.

By September 1955 the Nyeri Provincial Tea Board was able to note that farmers' enthusiasm for tea seemed to have been fully restored and it recommended that a tea factory be built in Mathira in the following year, followed by one in Othaya. In June 1956 a Central Province African Grown Tea Marketing Board was established, under which the Department of Agriculture retained responsibility for planting and cultivation standards and detailed instructions were promulgated. The board was made responsible for transporting green leaf to the factory from collecting centres, for factory construction, for tea manufacture and marketing. It was also responsible for servicing loans needed for nursery development, factory construction and operations. The factory at Ragati was opened in July 1957, even though there were only 377 acres of planted tea, most of it immature. But there was an overriding political compulsion to have a convincing demonstration of the potential of the new crop if it was to be taken up widely, and if farmers were to have confidence to look after their plots while the bushes matured for regular plucking. The factory-financing problem had been resolved by a loan of £15,000 from the African Land Development Board. I was able to visit the factory shortly before it opened and sent the following account back to Oxford: "Sitting on a stone beside the Ragati river, which tumbles like a Scotch burn, clear and bubbling, over lichened stones. Mount Kenya rises clear and beautiful in its lonely eminence from the far bank... It is hard to believe that this beautiful spot on the edge of the forest boundary should only a few months ago have been a place of fear and danger. Behind me on the hill stands the new tea factory, due to start manufacture this year of leaf supplied by upwards of 5,000 peasant farmers in the surrounding area, each to expand to half an acre of tea. I visited their neat holdings yesterday."[18]

The provincial administration, as well as the Agricultural Department, were insistent that all field staff should come under their control. As Gamble put it: "I do not agree that the Tea Board should have any staff in the field. I think the Department of Agriculture should have an Assistant Agricultural Officer (Tea) and that he should come directly under the Provincial Agricultural Officer... I do not agree that in the African Areas there should be any officer who is not a member of the department advising on tea planting, pruning and so on. This is a Departmental responsibility and as in the case of coffee, must always remain so if the quality of the tea shamba and the tea itself is to be kept up."[19] Tea from the Ragati factory was sold by the Central Province African Grown Tea Marketing Board; some was sent to the tea auctions in Nairobi for export through George Williamson; some was delivered to the Pool for the East African market; and some of the lower grades were disposed of locally by direct sales. Some of the farmers also started selling hand-made tea. These latter developments incurred the wrath of Brooke Bond, the managers of the Pool, who threatened to expel the Marketing Board, since "members are not allowed to sell in the local market in competition with Pool teas."[20] It was one thing for the Marketing Board to mend its ways, but the problem of sun-dried tea proved harder to eradicate.

The planting momentum was now increasing, although the contemporaneous pressure to complete land consolidation quickly was disruptive to planting programmes. For example, planting in Othaya division dropped from 34 acres in 1956 to under six in the following year, before increasing to 60 in 1957; in Mathira division, 81 acres were planted in 1957, but only 25 in 1958, again due to land consolidation.[21] At the end of 1957 the Central Province board looked forward to a fourfold increase in the tea acreage by 1960, on a sharply rising curve, and the question of financing tea factory construction became urgent.[22]

Enter the Commonwealth Development Corporation

The opening of Ragati tea factory in 1957 was a demonstration that smallholder tea could be satisfactorily grown and manufactured in Central Province. Many thousands of farmers had now embarked on tea growing, and the factory was receiving green leaf from up to 50 miles away in Othaya. A wave of enthusiasm for the new crop was developing and Roger Swynnerton was soon preoccupied with the financial implications. As he noted in June 1958, "Government is short of funds to put up factories in these areas and therefore there is a likelihood that tea development will be held up if outside sources of finance cannot be found."[23] He went on to envisage a model whereby a private company might finance a tea factory with political risk coverage from government, and either purchase green leaf outright for manufacture and sale, or else act as an agent on behalf of the provincial tea board. But it was also desirable, he thought, to make provision for smallholders to participate in the share capital of the tea factory in due course, perhaps financed by means of a cess on green leaf payments levied by the provincial tea board, acting on behalf of individual growers. The notion of direct shareholding by farmers in the factory companies was rejected as premature. This line of thinking became the basis of exploratory discussions with commercial tea interests and the regional controller of the Commonwealth Development Corporation over the following months, for which Gamble prepared a discussion paper that reviewed the development of African tea growing, the problems that had been overcome in establishing a satisfactory supervisory regime and options for associating private capital in factory development.[24]

Discussions were held with three tea plantation companies, James Finlay, Eastern Produce and George Williamson, with agents for tea machinery and with tea brokers. Brooke Bond was conspicuous by its absence, pleading pressure of other commitments. George Williamson had been the first company to make an agreement with the provincial tea board to purchase green leaf from up to 200 acres of smallholder tea for its Karirana factory, although only 37 had actually been planted in 1958. The major potential source of finance was identified to be CDC and its regional controller, Angus Lawrie, was drawn deeply into the evolving plans. In September, Swynnerton went to London to discuss with CDC the possibility of its direct involvement in the tea factory programme, following careful briefing by Lawrie. The reception was positive and in December the Kenya government made a formal application to CDC for both financial and managerial support.

The government had now refined its model for associating private capital. Individual factory companies would be established with an ownership structure consisting of 30 percent held by a private sector partner, who would be responsible for building the factory, for tea manufacture and sales, and 70 percent held by CDC, with a provision that it would be prepared to sell down a 30 percent stake to the provincial tea board acting on behalf of smallholders and perhaps other African investors. The tea board would have the power to raise a cess from growers to finance the stake. Thirteen new factories were envisaged over the ensuing ten years at a capital cost of £920,000. CDC's Executive Management Board agreed in principle to participate in the project and, as a first step, to consult with interested tea companies over mounting an investigative mission early in 1959. The Kenya Governor, Sir Evelyn Baring, joined the first exploratory meeting with the chairman of James Finlay, Sir James Jones, along with Lord Reith, chairman of CDC, and the general manager William Rendell. Baring was very well informed about the progress of land consolidation and smallholder tea development and the Finlay directors showed enthusiasm for becoming involved, but stipulated that they would only wish to build a factory near Kericho. A follow-up meeting in Nairobi a fortnight later showed that Swynnerton had given further thought to the issue of African ownership participation: the approach was still paternalistic in that the farmers' interest was to be an indirect one through the provincial tea board. But the growers would be made shareholders in the tea board and would receive dividends distributed by the factory company from its profits. Sir James Jones was pleased to see that there should be a means to secure farmers as shareholders in the factory company over time, but he was still concerned over the question of security of supply of green leaf and quality control in the light of the pattern of scattered individual plots and long transport hauls. Swynnerton and Gamble explained firmly that the consequence of land consolidation and enclosure ruled out any prospect of planting tea in blocks, which anyway had not proved popular in Othaya and Nyeri.[25] The concern over responsibility for leaf quality was met with something of a fudge: the Agricultural Department would maintain control over cultivation standards, the factory companies would have the right to reject substandard leaf and there would be "close liaison" between the two parties. A disappointment for Finlay was that the government's priority was tea development in Central Province and no factory for the Kericho area was envisaged until 1965–66.

The mission set up by the government was ready to start work at the end of March 1959. Five companies participated in its work: Eastern Produce, James Finlay, George Williamson, Dalgety (agent for tea machinery) and RE Smith (managing agent for Sotik Highlands), together with departmental officials and CDC, who had prime responsibility for the investigation as prospective lead financiers. The mission reported in April, with separate assessments of Central Province, Kericho and South Nyanza, Nandi Hills and North Nyanza.[26] One of the CDC members of the mission, Christopher Walton, was to play an important role in the future development of the smallholder sector. Over and above the official report, Walton and a colleague wrote an extensive confidential report for CDC and, furthermore, engaged in exploratory discussions with Swynnerton outside the framework of the mission.[27] Thus, there were several layers of evaluation of the mission's assignment.

It was set up ostensibly to assess the feasibility of financing tea factories to process tea from the smallholder sector. The mission was impressed by the high standard of cultivation on the farms it visited in Central Province and by the quality of leaf produced for the Ragati factory. It recognised that this was due largely to the control exercised by agricultural officers and the provincial administration and to the standards of supervision that had been established. In effect the mission acknowledged that Gamble had solved the problems of cultivation standards, transport and leaf quality that had plagued smallholder tea in India and Ceylon. CDC and the tea companies would have much preferred to follow a development model in which a tea factory was supported by a nucleus estate owned by the factory, and then to purchase green leaf from surrounding smallholders, so as to ensure basic viability. But, with one identified exception in the Nyambeni Hills in Meru district, population density, coupled with land consolidation and land enclosure, precluded this approach. It was acknowledged that this increased the risks of factory investment and, in an effort to mitigate them, a number of conditions were stipulated: that prior to investment there should be a planted acreage sufficient to support a minimum level of operation (500 acres), coupled with credible evidence of intention to increase the planted area to 1,200 acres for each factory. Furthermore, there should be supervision resources in place to support the development of the full unit. This produced a 'catch-22' situation. CDC and the companies would be reluctant to invest until there was the certainty of adequate tea to manufacture, coupled with a continuing strong supervisory regime, whereas the government was reluctant to continue encouraging tea planting and the expansion of its field administration without the certainty of factory facilities being in place. How could this potential stand-off be resolved?

The mission's proposal was that there should be a pragmatic and slower rate of tea planting, coupled with government assurances that the requisite acreages would be established. This meant that new development should fan out from existing factories at Ragati and company ones, so that there was a credible processing capacity until new factories were in production. This went against the plans of the Department of Agriculture, which wanted to start planting in a number of remote locations. There was a particular problem in Kiambu, where there was strong local pressure to expand quickly but no convenient factory nearby. There remained two worrisome issues. The first was whether the government would be able to devote sufficient manpower to maintain Gamble's standards of supervision and leaf quality at the rate of expansion now proposed. The mission was frank in its assessment that it had been unable to satisfy itself that adequate resources would be made available. The other concern was that a smallholder tea sector should be set up on a fully commercial basis, without hidden subsidies that would strain government finances, and capable of servicing loan capital. Farmers already had to purchase their tea seedlings from the nursery, but the price charged was less than half the cost of producing them. Furthermore, the supervisory regime provided by the agricultural department was a free service, although there were plans to start recovering costs through a cess on green leaf payments. On Gamble's model of one acre of tea per family holding (and a presumption of no hired labour), there was a prospective income of about £90 a year to which could be added other farm income. Development costs (purchase of seedlings and tools) were estimated at £9. This compared with an income of around £12 a year from farms without a cash crop.[28] Given that coffee was not a competing crop at altitudes over 6,000 feet, the mission concluded that there was an adequate financial incentive to grow tea; however, it was considered important to introduce a cess on green leaf payments without delay before a false picture was created.

The mission's conclusions on factory development on the evidence of current planting were that six factories would be justified on the following timetable (implying a decision to invest a year earlier):

- 1961: Othaya
- 1962: Mataara and Embu
- 1964: Kimulot and Kisii
- 1965: Nandi

This was half the number of factories envisaged by the Department of Agriculture. The mission had received widely differing estimates of capital costs and had been rather disconcerted by the heavy investment incurred by Brooke Bond and James Finlay in Kericho, but it eventually concluded that the smallholder factories could be built at an initial cost of £70,000 each, rising to £100,000 at full production of 1 million lbs of tea a year. CDC then calculated that at prevailing market prices of Shs3.50 per pound there should be a return on capital of 30 percent.

Finding willing commercial partners to invest in and manage the new factories proved difficult. James Finlay was seen as the strongest candidate, with impressive estates and management, and it was the only company taking seriously the potential of vegetative propagation; unfortunately, it was only prepared to invest in the

Kericho area, which would defer its involvement for several years. Eastern Produce showed considerable enthusiasm for investing in Central Province, despite all its field operations being in Nandi, and it also decided to partner CDC in its nucleus estate tea project at Nyambeni. George Williamson was favourably regarded as managers, but a problem was perceived over its willingness to commit capital and it stood aside from the initial commitments. Brooke Bond was a disappointment in that it declined to participate. Dalgety expressed interest in two factories.[29]

The official report of the mission did not press the issue of how future supervision of smallholder tea development was to be resourced, but it was clear that a serious problem had been identified. The mission had been triggered by Swynnerton's realisation that the financial requirement to build a dozen or so factories to process the tea grown by smallholders was far beyond the likely resources of the government. But Walton in his confidential report had now drawn attention to the much larger problem of how to manage such a major new agricultural enterprise. The report noted: "We have no reason to doubt Government's view that African smallholder tea growing can be a profitable business undertaking; we do question however whether Government has the commercial capacity to carry out the proposed programme. We also seriously doubt whether Government has the financial resources to maintain present standards of supervision on an increasing scale. In our view Government needs commercial and financial assistance."[30] The substance of informal discussions he had with Swynnerton while writing the report was the notion of setting up a tea development authority, supported by CDC, which would command the staff and financial resources to realise the potential of a large smallholder sector. Swynnerton wrote immediately to CDC that Walton's proposal "is of great interest to this Ministry" and requested that he submit detailed recommendations to the Ministry of Agriculture.[31]

CDC's Nairobi office was authorised to initiate discussions with government. To begin with there was uncertainty whether to focus solely on tea in Central Province, or perhaps just on the proposed

Nyambeni nucleus-estate scheme, or whether to include other cash crops and planned developments elsewhere in the country.[32] In September 1959 consideration had reached a point whereby the government issued a formal invitation to CDC to take part in a study to establish a new statutory authority to be responsible for the financing and management of smallholder tea development throughout Kenya, excluding the manufacturing stage, and ultimately to handle a number of other cash crops as well.[33] It was then found that the Ministry of Agriculture was not equipped to work on detailed plans for such an ambitious project and Walton was therefore assigned to the working party. This placed Walton (and CDC) in a unique position to assess the capabilities and capacity of the Kenya Government to implement its ambitions for smallholder cash crops. Over the next 18 months this had a profound effect on the shaping of the project and, what had started as a reasonably straightforward request for tea factory finance, drew CDC into devising an institution of formidable potential.

It was made clear to CDC that government wished the new statutory authority to assume full responsibility for the financing and direction of cash crop development, and that CDC and other parties would be appointed to the board in line with the usual policy of associating private sector interests. Reporting on a meeting with the acting director of agriculture, Leslie Brown, the CDC regional controller, Angus Lawrie, wrote: "(Brown) feels that this would be a very important consideration in winning the confidence of private enterprise to which Government looks for assistance in the financing and management of processing facilities." Lawrie continued:

> "The introduction of outside interests in such matters would be no new development in Kenya: the Board of Agriculture and, in fact, most statutory boards have for a long time included representatives of the farming community and of commerce. I would expect Government to continue this policy of integration of official and non-official in the composition of agricultural development and production boards."[34]

The government's initial view was that CDC would manage the authority on an agency basis, but Lawrie advised against this and thought that the authority's board should have full responsibility for managing the smallholder sector: developing tea nurseries and distributing planting material; maintaining farming standards (by employing unseconded agricultural officers); transporting green leaf to the factories; sale of green leaf; and the distribution of proceeds to growers. It would finance its operations by means of a cess on green leaf production.

The working party had no difficulty in endorsing the need for a new statutory authority to assume responsibility for the smallholder cash crop sector, but it recognised that the immediate priority was tea and framed its recommendations accordingly. It was realised that the programme needed the assurance of a long-term financial commitment outside the normal budgetary procedures of government; further, that the amounts required to realise the economic potential for tea growing far exceeded the government's available resources, so that a credible entity was needed to undertake external borrowing. Another consideration was that the project would require commercial direction that was outside the experience of government.

The working party recommended that the Department of Agriculture's responsibilities be narrowed to the supervision of field development, laying down cultivation practice, running the tea nurseries, research and the training of field staff, for which it would receive a fee, and that the other functions of the existing provincial tea boards run by the department would be absorbed into the new authority. It proposed accordingly that the new authority be empowered to borrow and itself to make loans in kind in the form of services and supplies to farmers. It was also proposed that the authority would take over and run the Ragati factory, leaving all new factories to be separately constituted on the basis already discussed by CDC and the tea companies. As to the development programme, it was recommended that there should be seven new factory units, with five of them in Central Province, including the Nyambeni nucleus-estate project that Eastern Produce was keen to adopt, and also a factory at South Imenti in Meru district, so that each of the tea-growing districts should have access to a tea factory.[35] The proposed factory in Nandi was dropped. All this implied planting nearly 7,500 acres by 1967, in addition to the 1,500 acres already planted, thus making a project total of 9,000 acres. On the basis of development costs of £120 per acre for the new planting, total expenditure would amount to £935,000, of which £600,000 would need to be borrowed by the authority, after allowing for self-generated funds. The working party calculated that the borrowing could be recouped from farmers at the rate of £10 a year and that an overall cess of 12 cents a pound on green leaf (on a guide price of 40 cents a pound) would also cover supervision costs. This meant that a farmer could look to receive a net income of £63 a year from an acre of tea, compared with an estimated £12 a year average farm income in Central Province without a major cash crop.

Lawrie reported that the Council of Ministers accepted the report of the working party on 19 December 1959 and that he had met with the Ministers of Agriculture and Finance to discuss implementation. On the last day of the year, the Minister for Agriculture, Bruce McKenzie, wrote to invite CDC to participate in the project, with an indication that the government would be prepared to provide up to £125,000 of the required financing.[36]

Another working group was immediately set up to refine the financial estimates, and this further examination began to expose some of the difficulties with the whole enterprise. By following through some of the detail one can begin to appreciate the risks and concerns felt at the time, and also how the probing by CDC was of such significance to the shape of the eventual structure.[37]

The new working party concluded that there should be a 20 percent increase in the proposed extent of tea acreage under the authority, as well as an acceleration of the planting programme, which would bring the total area up to 11,000 acres by 1967. The justification was to bring within the ambit of the authority all smallholder tea planting in the country, and not just the seven factory schemes previously identified, so that it would become a national smallholder authority. More careful estimation of recurrent costs led to a doubling of the proposed cess to ten cents per pound. Total development expenditure to be capitalised was estimated at £1.0 million (at a lower cost per acre than previously), and it was decided to keep the capital cess at seven cents per pound on the basis of a more extended repayment period and, more importantly, due to a more favourable view of yield prospects. Having looked at results at Ragati, it was felt safe to raise the assumed yield from 1,000 lb per acre to 1,250 lb per acre. A total cess on green leaf payments of 17 cents per pound implied a reduction of prospective income to the farmer from £63 to £52 on the old model but, with the new perception of higher yields, the revised estimate for each farmer's net annual income was £65.

As to financing requirements, the government agreed to increase its cash contribution from £125,000 to £235,000 so as to include improvements to rural roads over which the leaf-collecting lorries would travel, which left an estimated need for external financing of £1.1 million (including deferred interest). CDC was requested to provide this sum, with a request for an urgent decision in view of the momentum of development already under way. In macro terms, the project was projected to generate net incomes to farmers aggregating £715,000 at maturity, for a capital borrowing of £1.1 million, and the enthusiasm of the government for the scheme was understandable. In an astute move, the government let it be known that it proposed to appoint CDC's regional controller, Angus Lawrie, himself a leading business figure in Nairobi, as first chairman of the smallholder authority; and that it was making arrangements to appoint one of its most vigorous and effective district commissioners, Douglas Penwill, as its chief executive. Meanwhile, Gamble, who had reached retirement age in the Department of Agriculture, was to be appointed chief technical officer.

CDC still had a number of concerns over the project. The underlying worry was that the government had committed itself to a national programme of smallholder tea development before a proper management track record had been established, other than the demonstration of good standards of cultivation. By commencing planting in every district that was capable of growing tea, a political momentum had been created that was leading to ever-increasing acreages every time the scheme was assessed. CDC realised the impracticability of insisting on a pilot project at this stage as it would have preferred; however, it decided it could limit its financial risk by stipulating a slower pace of development. This was to be effected by reducing the planned acreage supporting each factory from 1,200 to 1,000, leaving a reduced target of 9,035 acres by 1967, and in this way the maximum financing that would be required from CDC would fall from £1.1 million to £0.9 million. Part of the concern was due to the fact that CDC was seen as the sole external financier and it expressed the wish that the World Bank be invited to come into the scheme. Meantime, CDC had experienced difficulty in securing Treasury clearance to participate in the project, on the grounds that it would be doing business that was more appropriate for a bank. This was the notorious 'finance house' issue that plagued CDC in its early years: the notion that it was stealing business from the City on the back of cheap government financing. As the narrative will have illustrated, this was scarcely traditional banking business as then conceived, and Kenya's governor lobbied energetically on CDC's behalf.[38]

A major area of concern related to the magnitude of the administrative task facing the new authority for a project involving many thousands of independent growers harvesting throughout the year, transporting green leaf over a rural road network to independently managed factories, maintaining accounting records for every farmer and making cash payments to them. Walton had formed a poor view of Gamble's capability in this area and of the way that the precursor provincial tea marketing boards had been run and their lack of commercial rigour and he minuted: "From what I have seen and learned of the existing development during the last 12 months I can say categorically that I would not support any CDC involvement in tea factories on the basis of the present organisation of the smallholder tea development."[39] One of the difficulties was the dispersal of responsibilities between the agricultural administration in the field, the factory administration by the companies and the new authority standing in between the two parties. The field administration relating to tea was inextricably mixed up with other duties of the agricultural department and with the elements of subsidy and partial cost recovery that had evolved over several years. The factory administration had still to be created, apart from the precedents of Ragati and the company-bought leaf schemes, where there was a variety of practice. Small wonder that Rendell, CDC's General Manager, was anxious about the practicalities and worried that the new authority was being tasked with too large a mandate.[40] The resolution was that the Agricultural Department would only be responsible for field development as agent of the Authority, while the new Authority would be responsible for leaf harvesting and transport to the factories and the all headquarters administration; it would not itself undertake factory development.

Finally, there was concern over the degree of independence of the board of the new authority, and especially how far it would be subject to political interference through directions from the minister of agriculture. Assurances had been given that the minister would only be able to give directions of a general nature, and the proposal to appoint Lawrie as chairman was clearly intended to strengthen the impression of an independent board. Nevertheless, there would be government appointees on the board, as well as grower representation. Much would depend on the early conduct and reputation established by the new board.

CDC decided at the end of July 1960 to go ahead in terms of the framework recommended by the working party, except for the stipulation of the reduced acreage and financial commitment, if it was to be the sole source of external finance. However, if the World Bank came into the project CDC would welcome the full programme being undertaken, and the Kenya government promptly commenced overtures. The Special Crops Development Authority (SCDA), as it was called, was established in September 1960, with Lawrie as chairman and Walton as a member of the board, and took over the assets and liabilities of the two provincial tea marketing boards.[41] Penwill was appointed chief executive and Gamble chief technical officer. CDC drew up a supervision agreement appointing the Director of Agriculture as agent with the authority to supervise and train farmers. The Minister for Agriculture, McKenzie, made a statement to Legislative Council on 16 December and there was a debate, with tribute to the contribution made by CDC. A memorandum from head office captured the mood: "Our view is that CDC participation in this new scheme is important enough to merit a flourish of trumpets."

Special Crops Development Authority

In 1959, the year prior to the formation of the SCDA, there were 1,246 acres of smallholder tea in Central Province, 300 in Nyanza Province and 25 in Nandi – a total of 1,572 acres.[42] The SCDA was now charged with planting 7,463 acres of smallholder tea, the bulk of it in the ensuing four years. It would be a formidable management challenge to absorb the operations of the two provincial tea boards and establish a new working relationship with the Agriculture Department; to organise the supply of millions of stumps for the planting programme; to arrange transport for delivering green leaf to the factories; to set up accounting systems for payments to farmers; and generally to implement a transition from a subsidised government extension service to tea farmers into an independent, commercially viable undertaking that would be capable of attracting and servicing large amounts of loan capital, as well as negotiating a series of factory agreements with the tea industry.[43]

The two African Tea Marketing Boards had been responsible for establishing tea nurseries and the provision of plants to farmers, and for purchasing green leaf and conveying it to the factory. In Nyanza the leaf was sold to tea companies; in Central Province the Marketing Board owned the factory at Ragati. The two boards had different payment systems and were more concerned with promoting tea growing than with balancing their books. The SCDA took over their assets and borrowings and rationalised procedures, setting up records for every farmer's leaf deliveries, revenue earned, deductions and indebtedness, in quarterly statements for some 9,000 accounts initially, but with a capacity for 20,000 accounts.

The board made plans to build up a field staff of its own to manage leaf collection and transport, comprising eight officers and 234 junior staff. It also had to arrange with the Department of Agriculture the recruitment of ten tea officers, 23 instructors and 94 assistants, all of whom would be on the strength of the department but with their cost reimbursed by the SCDA.

The tea nurseries had the task of growing tea plants from seed and maturing them for two years. Seed imports from India and Sri Lanka were banned on disease-control grounds and had to be sourced from within Africa, including Tanganyika and Congo. With a planting density of 3,000 plants per acre, the annual requirement was for over 5 million stumps. The SCDA found that it needed to purchase some 40 percent of its needs from the nurseries of the tea companies. As Penwill noted, "the movement of the stumps resembled the planning for the transport of an armoured division in war." [44]After allowing for these commercial purchases, an average price of 30 cents per stump was fixed and a set of tables was drawn up to determine a farmer's investment cost, depending upon whether he paid the basic price of six cents, or a higher fraction, which determined the amount of debt to be recovered through a capital levy.

The Kenya government had undertaken to upgrade essential tea roads, but the SCDA was responsible for organising leaf transport to factories. Farmers had to bring their plucked leaf to buying centres, which were spaced out along the tea roads at three-mile intervals, with the aim that no one would have to travel more than two miles. On buying days each farmer's harvest was turned out onto hessian shelves for examination by the leaf inspector, who insisted on a standard of "two leaves and a bud" and rejected coarse plucking. This served to establish the high reputation of smallholder tea from the first Ragati factory. On being weighed, a receipt was issued to the farmer and this documentation then formed the basis for the payment schedules – initially a payment of 40 cents per pound. There was provision for a second payment at the end of the season, depending upon actual prices realised.

A calculation was made of the running costs of the SCDA when all the planted tea had reached maturity by 1971, which formed the basis of a levy of ten cents per pound to offset running costs. There was also the matter of recovering development and financing costs – estimated at £1.5 million by that date – where it was determined that a capital levy of seven cents per pound be imposed on all tea planted after 1960, apportioned on the number of tea bushes planted as well as on a quantum of green leaf. The calculation was designed to recover capital costs by 1984. The payment of 40 cents per pound for green leaf was therefore subject to a deduction of 17 cents to recoup the SCDA's recurrent and development costs. On the assumption of a yield of 1,000 lb of made tea per acre (which was already looking conservative), this indicated that a farmer could expect an income at 1960 tea prices (which were approaching Shs4 per pound) of £51.75 per acre, before anticipating any end-of-season supplementary payment.

An early priority was to set up a network of tea-grower committees at location, district and provincial levels in order to provide a channel of communication and for the airing of concerns. This quickly proved its value in dealing with the introduction of cess charges and over the adjustment of planting allocations.

The foregoing details help to establish how much attention was given to setting up a structure for smallholder tea that was self-sufficient and not dependent upon the provision of free government services (albeit still with a reliance on undertakings to maintain effective tea roads). This was to establish financial credibility so that the SCDA would be able to secure and service loans from CDC for the First Plan to bring the smallholder tea area up to 9,000 acres, and even more so for future expansion. Walton returned to CDC in London in mid-1961 satisfied that an independent authority operating on commercial lines had been established. In a valedictory letter he identified two possible areas of future difficulty.[45] He was concerned that tea farmers were the only ones being made to pay directly for the cost of their supervisory services, and expressed the hope that other crops would be brought within the SCDA's orbit. He was more concerned that there might be rebellion against the strict controls that the SCDA imposed to ensure quality, and that contemporaneous preparations for political independence could lead to loss of authority. This aspect was debated at a meeting of the board, when the industry member, Sir Colin Campbell from James Finlay, argued that the authority was being too strict and that "there should be a gradual relaxation of the stringency of control in regard to pruning methods, plucking etc."[46] This was countered strongly by Penwill: "Unless standards of cultivation were laid down and enforced in the early years of development, in the opinion of the officers the scheme would be a financial failure."

CDC's decision at the end of 1960 to cut back the plan of the working party by reducing its financial commitment by 18 percent to £0.9 million, and reducing the area of new tea planting by 1,900 acres to 7,436 acres, created a severe problem for Penwill on the ground in the following year. The tea nurseries were geared to the original programme; more importantly, there was strong demand from farmers to plant the new crop, and Penwill advised that it would be politically unwise to frustrate it.[47] Only in Embu was there lack of enthusiasm for tea growing because this district was also suitable for the better-known coffee. There was also a hold-up in Fort Hall due to an embarrassing need to remeasure much of the land consolidation in the district.[48] Penwill's other major consideration was to avoid sacrificing one of the new factory units, or to prejudice their success by having less than 1,000 acres of planned tea to supply them. His short-term solution was to ensure the original planned rate of development in 1961 in the districts that were keen to grow tea, by reallocation from Embu and Fort Hall and from districts supplying existing commercial factories.[49] His more radical proposal was that preparations should commence immediately for a Second Plan for the five years to 1968, in order to restore the cutback and to meet the enthusiasm for the new crop. He proposed an additional 4,550 acres, with new factory units in Fort Hall, Kiambu, Kericho and Kisii. This implied substantial new financing of £1.2 million. The SCDA board endorsed the plan at the end of the year and even went on to enlarge it by adding 1,200 acres (and a factory) in Nandi, which had only lately shown interest in tea planting. This raised the Second Plan planting target to 7,700 acres.

At this point there was no financing in place, but there were two encouraging indications. In 1961 the World Bank agreed to send an economic survey mission to Kenya, two years before the country's independence, to review its economic potential and to make recommendations for its development planning. This was a strong indication that financial support from this source would be forthcoming. Also, in anticipation of independence, the West German government began to assess development aid prospects and was particularly interested in the SCDA and tea. With the approval of CDC, details of the smallholder programme were provided to the German authorities and, early in 1962, the SCDA learned that a loan would be provided to the Kenya government to enable restoration of the 1,900 acres that had been cut from the original programme. At the same time, details of the Second Plan for an additional 7,500 acres were given to the World Bank mission. In response, the SCDA was asked to estimate the optimum programme that it could handle in the five years to 1968, to which the answer was 12,315 acres and five more tea factories. The enlarged total was duly endorsed in the mission's report.[50] The World Bank and CDC were to be major participants in subsequent developments. The latter's association was reinforced by a more intimate association: on retiring as Governor of Kenya at the end of 1960, Sir Evelyn Baring was appointed chairman of CDC and then, in 1962, on Swynnerton's retirement as director of agriculture, he joined CDC as its agricultural adviser and was closely involved in its further commitments to the smallholder tea sector.

The increasing momentum of field development required a response on the manufacturing side, with decisions required in 1962 on the first two new factories in Nyeri (Chinga) and Kiambu (Mataara) to go into production in the following year. A revolution in the technology of tea manufacture was in process at this time, and CDC and the SCDA had to make up their minds whether to play safe with the old technology (rolling, fermenting, drying) or to adopt the new method (cut–tear–curl) for the smallholder factories. At the time, only George Williamson of the plantation companies had made a commitment to the new method, and it appeared there was no price penalty for tea manufactured in this way. Capital costs, on the other hand, promised to be significantly lower because of reduced factory space. It was decided to go with the new technology. CDC's partners in the first two factories were a London merchant bank that had significant interests in Kenyan agriculture, Arbuthnot Latham (whose chairman was on the CDC board), and Dalgety & New Zealand Loan, who were active as tea brokers. The companies would provide loans to finance the SCDA's one-third share of the capital cost, with Arbuthnot Latham appointed as factory manager on a 2 percent commission for 18 years (the period of the loan) and Dalgety as brokers to the companies. The equity capital of the factory companies was to be shared equally by CDC and the SCDA and provision was made for the issue of shares to farmers at a future date up to a level of 49 percent. The agreement provided that the factory company would purchase green leaf from the SCDA at a price that allowed for operating costs, reserves and a margin for dividends of up to 8 percent. The factory boards would include representation from CDC, the SCDA, government and growers.[51]

The SCDA now ran into an unexpected problem that posed a serious threat to its financial viability for about three years. Towards the end of 1961 it was noticed that green leaf deliveries were falling off in locations in Central Province and that hand-made sun-dried tea was being offered for sale in Othaya and Karatina markets on an increasing scale. A more comprehensive survey early in the following year established that sun-dried tea was on sale in 33 markets in the province and that it was also being 'exported' to markets in the Rift Valley on an organised basis.[52] The reaction of Penwill and the SCDA board illustrates well the authoritarian approach to the management of the smallholder sector. Little interest was displayed in the possible reasons for this sudden appearance of an informal tea marketing network: that perhaps the cost of tea sold through the Pool managed by Brooke Bond was beyond the pockets of many poor consumers; that the quarterly payment to growers for green leaf was unattractive in comparison to the immediate opportunity to realise cash for sun-dried tea; and that the strict plucking standard imposed by the SCDA, when combined with infrequent leaf collection schedules, left an element of rejected leaf growth (the 'third' leaf), for which an economic opportunity was discovered to turn it into ready cash. Penwill's approach was to use the authority of the administration and of the Tea Rules (enforceable through the native courts) to maintain discipline among farmers. Barazas were held throughout the province and all the tea committees were pressed into service to condemn the diversion of leaf away from tea factories, and to stress the importance of producing a high-quality product. By mid-year, police powers were in place to confiscate sun-dried tea and to prevent its sale and movement, and prosecutions commenced.

The threat to the viability of the SCDA was a real one, for two reasons. In the first place, green leaf that was diverted to this quickly thriving market meant that the SCDA was unable to collect its ten cents per pound cess to meet its operational costs, or its seven cents per pound cess to defray borrowing costs. Penwill estimated that Ragati lost between 40 and 50 percent of green leaf input in 1961–62 as a result of the diversion to sun-dried tea, and warned that the financial foundations of the organisation were being undermined.[53] The other threat was to the plans for new factories, which were premised on each factory receiving all the green leaf from its catchment area – especially in the early years before they were operating at full capacity. The old fear of the tea companies that growers would withhold green leaf seemed in danger of being realised, from an unexpected cause. Penwill was fearful that CDC and other investors would refuse to proceed with the factory programme if the SCDA was not seen to be resolute in stamping out the alternative market. There was also a more general reputational concern that was felt keenly by Penwill and Gamble: much had been staked on the assertion that smallholder tea growing could be established under a disciplined regime of quality production, yet disorder now threatened. In Limuru, estates were being raided for green leaf for the new market (yet another original fear of the industry). More widely, there was a breakdown of plucking discipline: 'two leaves and a bud' might be picked for the Ragati factory, but then a third or more leaves would be collected for sun-dried preparation. (The fact that tea companies were harvesting coarser plucking on a regular basis in Kericho was an embarrassment to be overlooked.) Finally, there was the threat to the orderly structure of the domestic tea market, whereby all estate producers undertook not to market tea separately and only to sell production through the Pool managed by Brooke Bond. Sun-dried tea was expressly banned by Tea Board rules.

Penwill did not hesitate to enlist political support, including that of future president Jomo Kenyatta, as well as arranging for tougher tea rules: the selling of sun-dried tea had been made an offence which could be heard by African courts, and the police were given the power to confiscate, and prevent the manufacture, sale and movement of sun-dried tea.[54] The problem continued to worry the SCDA into 1964, although on a diminishing scale, partly as a result of the efforts to stigmatise sun-dried tea and prosecute offenders, but also through improvements in the leaf collection services to farmers and the fortunate impact of good tea prices being obtained for orthodox tea.

Kenya was set to gain its constitutional independence in December 1963. Amongst all its other preparations the government was anxious to formalise the financing of the Second Plan before that date, as there would otherwise be procedural delays involving its formal accession to the United Nations and its access to the World Bank. On the ground, meantime, there was mounting demand from farmers to grow tea, so that it was important politically to maintain the momentum of development. For CDC, as the principal prospective financing partner of the World Bank, there were challenging issues. The chairman of the SCDA, Angus Lawrie, was CDC's regional controller and was due to retire during the year, as were Penwill and Gamble, so that future management was a major concern, especially as there were rumours that the board would cease to be autonomous from government.[55] The lending quirks of the International Bank for Reconstruction and Development (IBRD) had also to be taken into account: normally, it would only finance the import content of a project and the import content of the Second Plan was put at £1.8 million out of £2.5 million; did this imply that CDC would be asked to finance the £0.7 million of local costs?[56] Another factor was the increasing accounting complexity of keeping the records of the First and Second Plans distinct, given the overlapping impact on the ground. In July CDC was formally invited to participate in the Second Plan and it cooperated closely with the World Bank mission that visited Kenya in September, leading to a decision in December to negotiate its detailed participation.[57] A further enlargement of the planting programme to 14,399 acres by 1969/70 was approved and to widen its geographical coverage.

The SCDA's achievement over the four years had been impressive. By 1963 there were 18,000 smallholder farmers with an average of 0.45 acres of tea each, cultivated to a high standard, and an effective supervisory structure had been put in place with the crucial assistance of CDC. This had won the confidence of the West German development agency KFW for it to participate in financing the field programme. More importantly, it had won the support of the World Bank to become involved in the Second Plan for a large extension of smallholder tea growing from 11,000 to 25,000 acres and an increase in the number of participating farmers to 30,000. The Second Plan would also require increasing the number of tea factories from seven to 17 and a financing model had been established for the new factory companies, with CDC as the principal investor alongside the SCDA – which itself would obtain loan finance from the private companies managing the factories. The first four firms to commit were James Finlay, George Williamson, Dalgety and Arbuthnot Latham. A distinctive feature of these ambitious plans was that thousands of smallholder farmers were being gathered into an institutional structure that was designed to service and repay the total financial investment and meet running costs, all from the proceeds of their tea cash crop. It was this that enabled external loans to be raised from IBRD, CDC, KFW and tea companies. This had not been the pattern with the earlier introduction of cash crops such as coffee and pyrethrum.

The only significant funding obligation for the Kenya government was to provide and maintain tea roads, which became a significant focus of concern. However, the popularity of this new source of income for high-altitude farmers was such that government support for smallholder tea was not in doubt. Indeed, Finance Minister Gichuru had written to CDC in November 1963 underlining the importance he attached to securing their commitment to the Second Plan and offering to fly to London or Washington if this would help with negotiations.[58]

Everyone was now onside, and this consensus included agreement to merge the First and Second Plans into an overall programme to achieve 25,500 acres of smallholder tea by 1970, which would eventually require 17 owned factories, but only seven in the plan period. On 2 January 1964 the name of the authority was changed to the Kenya Tea Development Authority, in recognition that it had more than enough to do with one special crop. In negotiations in Washington early in 1964 the World Bank decided to dissociate from participation in factory financing, notwithstanding its preference for financing import content, because of the procedural complexities involved in negotiating so many factory agreements.[59] CDC accordingly agreed to provide 50 percent of the factory finance up to a ceiling of £750,000 and the World Bank, through its International Development Association affiliate, agreed to provide loans of £1 million for field development. In the First Plan, CDC was the principal external source of finance, both for field development and factories. In the Second Plan, CDC still had the largest financial commitment, but the World Bank was now seriously engaged as well. However, the tea farmers themselves were committed to providing nearly 20 percent of field costs directly through the development cess, as well as ultimately repaying the loan capital. The financing of the combined First and Second Plans is set out in the table overleaf.[60]

Implementation of the enlarged planting programme was accelerated over the five years and by the close of the 1967–68 planting season the target had been exceeded to reach 15,799 acres, or 27,100 acres of smallholder tea overall, with an average holding of 0.71 acres of tea per farm.[61]

Financing of First and Second Plans

	£'000	£'000
Field development		
IDA	1,000	
CDC	900	
KFW	212	
KTDA (cess revenue)	1,941	
Kenya government	47	
Total		4,100
Factory development		
CDC	1,230	
Companies	740	
Kenya government	350	
Total		2,320
Government infrastructure		
Housing	90	
Roads	1,550	
Total		1,640
Grand total		8,060

SCDA Tea Planting Programmes

	Acres	
Already planted by 1960 (revised to 1,565)	1,572	
Working party plan	9,363	
Total planned acreage		10,935

First Plan

	Acres	
CDC	7,463	
KFW	1,900	
SCDA uplift	173	
	9,536	
A. Revised total acreage (including earlier planting)		11,108

Second Plan:

	Acres	
SCDA initial proposal	5,800	
IBRD uplift	6,515	
SCDA uplift	2,084	
B. Planned acreage		14,399

Total A+B   **25,507**

NOTES TO CHAPTER 10

[1] MP/African Tea/1/1, Letter to Sir Frank Stockdale, agricultural adviser to the Colonial Secretary, July 1939.

MP/African Tea/1/1, Letter to the Governor, 3 July 1946.

[3] MP/African Tea/1/1/, Senior agricultural officer, Central Province to Director of Agriculture, 20 April 1946.

[4] MP/African Tea/1/1/, Memorandum to the Member for Agriculture, 15 November 1947.

[5] MP/African Tea/1/1, Director of Agriculture to KTGA, 29 June 1948.

[6] MP/African Tea 1/1, KTGA minutes, 29 October 1948.

[7] MP/African Tea 1/1. The episode is rehearsed in a memorandum by Gamble dated 19 January 1953.

[8] *Report on Visits to India, Malaya and Ceylon with some notes for the guidance of tea planters in Kenya*, G. Gamble, 1951, Government Printer, Nairobi.

[9] MP/African Tea/1/1, *Tea In Central Province*, November 1950.

[10] There is a large literature on the Mau Mau rebellion and its relationship to the politics of

independence. To mention but one book, *Economic and Social Origins of Mau Mau*, David Throup, James Currey, 1987.

[11] MP/African Tea/1/1, African Tea Board minutes, 24 September 1953.

[12] MP/African Tea 1/1, Memorandum to Director of Agriculture, 5 March 1954.

[13] *A Plan to Intensify the Development of African Agriculture in Kenya*, 1954, Government Printer, Nairobi, paragraphs 37 and 38. Roger Swynnerton, later Sir Roger, became Permanent Secretary of the Ministry of Agriculture. After retirement he joined CDC as its agricultural adviser.

[14] MP/African Tea/1/1, *Tea Cultivation by Africans*, memorandum to Tea Board by Director of Agriculture, 12 June 1954. It was based on a detailed proposal prepared by the district agricultural officer.

[15] MP/African Tea 1/1, KTGA reply 12 July 1954.

[16] MP/African Tea 1/1/, Report of 23 August 1955

[17] MP/African Tea/1/1, Note by Gamble, April 1955

[18] Part of a letter to Margery Perham, Senior Fellow of Nuffield College, Oxford.

[19] MP/African Tea/1/2, Central Province Tea Board minutes, Gamble to Director of Agriculture, 25 July 1957.

[20] MP/African Tea/1/2, ATGA letter to the Central Province African Grown Tea Marketing Board, 29 September 1958

[21] MP/African Tea/1/4, *Expansion of Tea Planting in the Central Province*, G Gamble mimeo, 7 August 1958.

[22] MP/African Tea/1/4 The Central Province Tea Board minute of 23 December 1957 gave the following acreage projection:

1955	1956	1957	1958	1960	1962
181	371	491	740	1,940	3,140

[23] MP/CDC/1, *Participation by Private Capital,* memorandum, June 1958.

[24] MP/African Tea/1/4, *Expansion of Tea Planting*, op. cit.

[25] MP/CDC/1, Note of meeting held in Ministry of Agriculture, Nairobi, 9 February 1959.

[26] MP/CDC/1, Report of mission investigating methods of processing smallholder tea production in the Central Province of Kenya, with separate supporting reports on the other two areas, 13 April 1959.

[27] MP/CDC/1, *Confidential Report on Visit to Kenya*, 13 May 1959

[28] It was estimated that one acre would yield 4,500 lb of green leaf when mature and that, at current prices, the farmer would receive 40 cents per lb.

[29] MP/CDC/1, Minutes of meeting with tea companies, 2 June 1960.

[30] MP/CDC/1, *Confidential Report*, op. cit.

[31] MP/CDC/1, Swynnerton to Regional Controller CDC, undated April 1959.

[32] MP/CDC/1 *Notes on Proposed Smallholder Development Authority*, CDC, Nairobi. *Financing of African Tea Cultivation in Kenya*, CDC Executive Management Board 194/59, 19 June 1959. *Nyambeni Smallholder Tea Scheme*, Walton, 30 July 1959.

[33] MP/CDC/1, Ministry of Agriculture to CDC, 14 September 1959.

[34] MP/CDC/1, Regional Controller to London, 22 October 1959.

[35] In Central Province: Kiambu, Othaya, Embu, Meru and Nyambeni. West of the Rift Valley: Kericho and Kisii.

[36] MP/CDC/1, Minister of Agriculture to CDC, 31 December 1959.

[37] See especially MP/CDC/1, *Financing of African Tea Cultivation in Kenya*, CDC, BP16/60, 11 February 1960. *Kenya: Smallholder Development Authority*, CDC BP 60/60, 28 July 1960.

[38] This issue is considered more fully in the author's book *The Development Business: A History of the Commonwealth Development Corporation*, Palgrave, 2001.

[39] MP/CDC/1, Note by Walton to General Manager, May 1960

[40] MP/CDC/1, Note of a CDC head office meeting with Walton and staff, 18 July 1960.

[41] Legal Notice 458 of 29 September 1960, established the SCDA as a Local Land Development Board under the Agriculture Ordinance. It was then re-established, following amendment of the Ordinance, under the Agriculture (Special Crops Development) Order 1961 by Legal Notice 243 of 21 April 1961.

[42] Later corrected to 1,565 acres on re-survey.

[43] MP/CDC/2, The Special Crops Authority – Annual Report, 1961.

[44] MP/CDC/2, SCDA Annual Report 1961

[45] MP/CDC/2, CH Walton to controller of operations, London, CDC, 4 July 1961.

[46] MP/CDC/2, SCDA Board minutes, 20 January 1962.

[47] MP/CDC/2, *Future Planting Programme – Outline Proposals*, SCDA, 8 May 1961.

[48] *Land Reform in the Kikuyu Country*, MPK Sorrenson, OUP, 1967.

[49] Eight commercial factories had arrangements to process green leaf from smallholders. CDC/EMB 308/63 of 12 December 1963.

[50] *The Economic Development of Kenya*, IBRD, John Hopkins Press, 1963, pp.121–3.

[51] MP/CDC/2, *Kenya: African Tea Factories*, CDC, BP 19/62, 8 March 1962.

[52] MP/CDC/2, SCDA Board minute, 26 January 1962. Also circular to specialist tea officers, 'Sun Dried Tea', 5 February 1962.

[53] MP/CDC/2, SCDA Board paper 61, 16 July 1962.

[54] MP/CDC/2. "Mr Penwill is conducting his campaign with considerable vigour. He is enlisting the support of leading political figures, amongst them Mr Jomo Kenyatta, whom he is going to see next week." Letter from CDC Nairobi office to London, 27 June 1962.

[55] MP/CDC/2, Regional Controller's report for September/October 1963.

[56] MP/CDC/2, Board paper 52/63 of 1 August 1963.

[57] MP/CDC/2, Board paper 98/63 of 12 December 1963.

[58] MP/CDC/2, Minister of Finance to Regional Controller, 5 November 1963.

[59] MP/CDC/2, Board paper 27/64, 12 March 1964.

[60] MP/Lib/19, *The Operations and Development Plans of the Kenya Tea Development Authority*, Nairobi, 1964, pp.22–3.

[61] MP/CDC/2, Brief for Chairman, 18 September 1969.

11

THE KENYA TEA
DEVELOPMENT AUTHORITY

Implementing the First Three Plans

KTDA published its first report on its operations in December
1964.[1] It had been established as a corporate statutory body by Legal
Notice No 42 and was responsible to the Minister for Agriculture,
who appointed its chairman. Provision was made for grower representation, alongside the Tea Board, CDC, the chief executive and
others. Its remit was to promote tea development in designated
areas (the former African land units) and it was armed with powers
to make regulations (and to penalise non-compliance), to levy
cesses on growers and to control marketing. These powers enabled
KTDA to designate where tea could be grown (the 'tea line'), to
stipulate that tea could only be grown on freehold, consolidated
farms with an acreage limitation (one acre initially) and that
farmers must obtain their planting stock from KTDA and follow its
regulations for planting, cultivation and harvesting. In order that
KTDA's finances could be kept separate from those of the factories,
the latter were to be established as individual companies. KTDA
would purchase green leaf harvested by the farmers and convey it
to a factory for processing and sale, and there was provision for a
possible second payment to farmers, depending on sale proceeds.

The cost of tea plants was to be recovered from farmers through a development cess over a long period, while the net running costs of the Authority were to be recovered from a revenue cess. These cesses were to be deducted from the green leaf payment. The financial model was of a self-financing enterprise, funded initially by loan capital, which provided for the servicing and repayment of borrowings from the World Bank and CDC out of farm income. The Kenya Government was only financially responsible for the provision and upgrading of roads in the tea zones, and for a small element of initial expenses.

At its inception, KTDA was responsible for completing the tea planting programme of the First Plan on 11,000 acres and involving over 20,000 farmers, and for implementing the Second Plan to add a further 14,000 acres commencing in 1965. KTDA was under the strong leadership of a former district commissioner, Dick Raynor, and, as indicated above, he had extensive powers to enforce high standards and discipline. Notwithstanding the challenge of implementing the two plans there was strong pressure on KTDA, both politically and in the countryside, to undertake further expansion. In February 1965, Raynor wrote to the Minister for Agriculture to outline proposals for a Third Plan for a further 23,000 acres of tea, which would get under way in 1968 – halfway through the Second Plan.[2] His assessment brimmed with confidence: "The smallholders have almost without exception cooperated to the full with the Authority, and have been keen to take the technical advice offered to them... the quality of the leaf which is being produced by the smallholder is extraordinarily high and this is being reflected in the consistently high prices being obtained for made tea... there is no doubt that the ability of smallholders to grow tea successfully has been abundantly proved." The only fly in the ointment was that leaf yields had been below expectation. Raynor placed on record the political pressures: "the demand for further expansion throughout the country is now insatiable. The growers themselves in all areas are crying out for more tea, and this cry is being taken up by political leaders at all levels." However, there were constraints on what could be achieved. First, the tea nurseries would have to be expanded to produce an increased volume of two-year-old plants to supply to farmers. Second, the ethos of the smallholder programme had been to establish high standards of husbandry, involving a large amount of supervision and control, which should not be compromised. KTDA was currently resourced for planting 3,000 acres per annum, and Raynor judged that this could be expanded to 5,000, once nursery stock had been increased. Third, and notwithstanding the political pressures, the additional tea acreage should essentially comprise increasing the size of existing holdings, rather than moving to new areas. This was necessary for administrative reasons, and for financial ones, as development in new areas was twice as costly as increasing existing holdings towards the target of one acre. Finally, Raynor requested that, before making any detailed allocations, there should be a special survey of all land in the country suitable for growing tea, to replace the more rough and ready rainfall zone demarcation of the Swynnerton Plan. This was agreed and the former chief agriculturalist of the Ministry of Agriculture, Leslie Brown, was retained to produce a comprehensive survey by the end of 1965 – the Brown Report.[3]

Brown set out to establish the total area of land that was suitable for growing tea, having regard to climate and soil conditions. Most of this was already densely occupied by farmers. In the existing tea-growing areas the report calculated that there were 896,000 acres of suitable tea land, occupied by 130,000 farmers; in addition, it identified a further 600,000 acres occupied by some 90,000 farmers that were also suitable for tea. The economic model derived from the Swynnerton Plan was that 10 percent of a farmer's acreage could be devoted to a cash crop, with the balance reserved for food supply. According to this methodology there was therefore the potential for 90,000 acres of tea in the existing areas and potentially a further 60,000 acres in new areas. It was recognised that, over time, many farmers would plant up a higher proportion of their land to tea. When set against the First and Second Plan target of 25,000 acres of tea grown by 33,000 farmers, and the initial target of the Third Plan of a further 23,000 acres to reach a national total of 48,000 acres, the notion that the Kenyan smallholder sector had a potential of 150,000 acres and over 200,000 growers was dramatic. The estate sector at this juncture contained some 45,000 acres of tea. It was clear that the Third Plan would be but a stage on a much longer journey.

Brown proposed a careful methodology for prioritising new planting allocations: to reflect ecological conditions; to recognise the importance of establishing larger individual plots (referred to as 'thickening up') in order to minimise development costs and reinforce experience; and to redress the earlier emphasis on Central Province development by recognising the potential of Kisii and Kericho. Each district was then carefully reviewed and scored. Brown went outside his brief in two important respects. As the former senior agriculturalist in the Department of Agriculture, he was well placed to comment on operational aspects of KTDA and did not hesitate to do so. He was critical of the tea officer cadre and its management; he wanted the tea nurseries expanded for the accelerated introduction of vegetative propagation; and he intervened on the controversial issue of the KTDA policy of restricting growers from purchasing planting stock from independent suppliers. His most trenchant observations struck at the heart of the smallholder strategy: estate cultivation provided more income and employment for a given area than smallholder subsistence farming, and he regretted that this insight had not been applied to the new settlement schemes. It followed that he was in favour of successful smallholders increasing the size of their tea acreage.

With regard to the tea estate sector, Brown noted that in Kericho, where 26,000 acres had been planted, there was potential for another 11,000 on company land and most of this had not yet been licensed for development. In Nandi the comparable figures were 12,000 acres planted and potential for a further 15,000.

Both KTDA and CDC thought that Brown was overoptimistic with regard to the tea planting potential. Nevertheless, the Third Plan was reformulated in the light of his report and the target was raised to 30,000 acres over five years, instead of six, with planting to commence in 1968–69 and finish in 1972–73.[4] This implied increasing the annual rate of planting from 3,000 to 6,000 acres (the minister was pressing for 10,000 acres a year). The estimated cost of development had doubled, largely in response to pressure to bring in new growers. On the manufacturing side, experience was pointing to increasing the size of individual tea factories to handle 3 million lbs of tea a year, instead of 2 million lbs, but the plan still envisaged a need for 20 new factories.

The debate over the balance to be struck between new growers and increasing the size of existing tea plots of established growers continued during the year. Financial analysis strongly favoured the latter since this reduced development costs and recurrent costs and hence the cess burden on farmers. KTDA's finances were already under strain as a result of lower yields than had been anticipated in the first two plans. There were also concerns – picked up in the Brown Report – over the standard of field supervision. However, there was a very positive development arising from the evolution in the tea nurseries. The major tea companies were all by now fully committed to the new technique of VP from bushes that had been specially identified as producing high yields of 3,000 lbs of tea per annum or better, as against the traditional 1,000 lbs. Not only did tea plants propagated by VP from selected clones, as the jargon went, offer the prospect of much larger green leaf harvests, but farmers could be taught to take cuttings themselves from 'mother' bushes, instead of buying two-year-old seedlings from the nurseries. This offered the prospect of a large reduction in expenditure on nurseries, as well as an acceleration of the development cycle. A decision was taken that the Third Plan would all be based on VP from selected clones, whereas the earlier assumption was that VP material would not be fully available until the fifth year. A reformulation of the Third Plan in mid-1966 was now premised on a new target of 36,000 acres, with the annual planting rate rising from 6,000 to 7,200 acres.[5]

KTDA's financial planning was predicated on being able, once again, to secure the backing of the World Bank and of CDC. The World Bank had been very supportive of Kenya's development plans, especially with regard to the small farm sector, and it was well aware of the preparatory work on the Third Plan. CDC was represented on the board of KTDA and was also monitoring developments closely, but without making any advance commitment. By early 1967, arrangements were being made for a formal World Bank appraisal mission later in the year and, in anticipation, CDC agreed to participate in it, with a view to investing in both field and factory development.[6] The World Bank made an important procedural decision to recognise that it was no longer dealing with a series of discrete plans, but rather was making periodic loans to an ongoing programme of tea development.[7]

Notwithstanding these positive steps, there were some more worrying aspects of KTDA's operations that indicated caution. Harvest yields were turning out much lower than anticipated, partly on account of slower maturing of tea plots to full yield, but also due to inexpert plucking and faults in the leaf collection system. In the year 1966–67 the yield shortfall against plan was as much as 25 percent. This meant that there was a significant reduction in KTDA's cess revenue, which was supposed to cover its recurrent costs, and the Authority was operating at a loss. This fell to be made good by the government in terms of the external loan agreements.[8] Another concern was the state of the roads in the tea-growing areas. The government had received a large IDA loan for their construction, but it had failed to provide the local authorities with funds for their maintenance. A solution was reached in 1966 whereby KTDA itself would undertake routine maintenance and be reimbursed by government twice a year.[9] There was irony in this solution in that Brooke Bond had done a similar deal in the 1920s over the maintenance of the road link from Kericho to the railhead at Lumbwa.[10] By the time of the appraisal mission in September 1967 these problems had been addressed and CDC was able to deliver a favourable verdict: "KTDA is an effective organisation which has won the confidence and cooperation of growers through its field staff and its district committees. Top management is efficient and there has been a successful transition from expatriates to Kenyans."[11]

The major innovation of the Third Plan was that, instead of providing farmers with two-year-old seedlings at the rate of 3,000 plants per acre, farmers were provided with 20 rooted VP mother bushes from which they were taught to take cuttings and grow them on themselves. A significant increase in supervisory staff was provided for. The Third Plan was also based on more cautious yield estimates, in the light of experience, with full harvest yields not being assumed until the seventh year. It was finalised on the basis of planting 35,000 acres over five years, commencing in 1968, which would bring the total planted tea acreage to 60,000 by 1972. There was an emphasis on increasing plot size in existing areas, but there was also to be a big expansion in Kisii and Kericho. The new programme would eventually require the construction of another 15 tea factories, on top of the 17 of the first two plans. All of them would be equipped to produce 2 million lbs of tea per annum, which would entail upgrading the capacity of earlier factories. These were long-term commitments, since only six factories were actually in production in 1967.

The Third Plan was costed at £14 million. The field development costs of £10 million were to be funded largely by cesses levied on the tea farmers, save for an element of £1.8 million comprising KTDA's accrued operating deficit, which was to be met by loans from the World Bank's affiliate the IDA and CDC (although it was ultimately recouped from the farmers in the form of an additional cess). The £4.3 million cost of new factories was to be met by loans to the factory companies from CDC and the managing companies, and retained earnings.

There was not an accepted methodology for estimating farm income from tea, nor was KTDA equipped to collect survey data from farmers. The calculation had its complications: harvest yields increased over six years as tea bushes matured, but they were also affected by plucking methods, the weather and, above all, by the planned switch to high-yielding VP plants; the automatic cess levied on harvested green leaf had a revenue element (ten cents per pound) and a capital element (seven cents per pound) – both were deductions from cash flow, but the capital element would gradually be eliminated; most income estimates were based only on the initial 40 cents per pound payment for green leaf and ignored the discretionary second payment (which was paid regularly) that added 25 percent or more to farmers' tea income; finally, none of the estimates took account of the cost of discretionary farm inputs, especially fertiliser. KTDA's 1964 inaugural publication gave a calculation for a mature tea holding, using the initial payment but ignoring the capital cess, to produce an estimated return of £61 an acre. In 1967 the CDC board was presented with a figure based on mature tea for the average size of tea holding amounting to an income of £50 after the revenue cess, equivalent to £67 an acre, but ignoring the second payment. In 1968 a new estimate was presented to the board, of £58 per acre on mature seedling tea and an expected income of £74 on mature VP tea.[12] The World Bank supervision mission to Kenya in September 1968 did not undertake any field visits and its report did not attempt to examine farm incomes. Using the methodology of the 1969 Working Party on Tea, which recognised paid labour costs and also the second payment, one can calculate that an acre of mature tea at that time (excluding the new high-yield varieties) would have yielded a net income of approximately £52.[13]

The early calculations overestimated harvest yields, thus overstating actual farm income, as was the effect of ignoring the capital levy; however, this was offset by omitting to take account of the important second payment. In their formal papers both CDC and KTDA gave very little attention to farmers' incomes, as compared with the detailed assessment of the overall financial position of KTDA and the position of CDC and the World Bank as lenders. Everyone was, of course, aware of the very strong popular demand to become a tea grower and of the importance of this new source of income to farmers in the tea-growing districts. Farmers appeared to take in their stride the low initial return from immature tea bushes and the burden of the cesses, in the expectation that yields and income would improve over time, as would the size of individual tea plots.

The devaluation of sterling in November 1967 did not affect the completion of tea planting under the Second Plan, but its potential impact on the Third Plan was a cause of great concern. Kenya did not follow the UK by altering the exchange rate of the shilling. World tea prices, led by its largest market, did not adjust upwards to offset the devaluation, with the result that Kenyan producers suffered a reduction of some 15 percent in their export earnings. The situation was exacerbated by the closure of the Suez Canal, as f.o.b. prices were lowered. The fear was that the reduction in earnings would discourage farmers from continuing with the development programme, and so attention was directed to possible offsetting measures. The government set up a working party which documented the problem in detail and recommended financial assistance to the smallholders (but not to the tea companies).[14] However, the World Bank appraisal mission in December 1968 took a hard-headed view "that farmers would continue to grow tea as this is, in most tea growing areas, the only cash crop and the opportunity cost of growing it is low."[15] This view prevailed.

The devaluation episode threw a revealing light on the smallholder programme. Its external financing through the World Bank and CDC, within a tightly defined framework, had the consequence that the tea farmer became directly responsible for servicing and repaying the foreign currency borrowings out of their harvest, including paying for the government extension services. The Kenya Treasury actually made a handsome return on the money (between the IDA lending rate and the rate it charged KTDA) and had eventual use of the funds for 30 years after KTDA had repaid its loans and before repayment to the IDA. This episode also brought to prominence the contrast between the regular initial payments for green leaf, which offset work inputs, and the lump sum second payment at the end of the season, which was the true reward to farmers from their crop.

The planting programme of the Third Plan was implemented much faster than anticipated. Notwithstanding the concerns over the impact of sterling devaluation on farm incomes and motivation to continue planting tea, 5,900 acres were planted in the first season, while favourable weather conditions facilitated the planting of 6,600 acres in the second year. In this year, not only was the crop larger but prices had also improved as a result of labour unrest in India and the holding of international discussions on output quotas. The average yield rose to 1,122 lbs of made tea per acre and the second 'bonus' payment averaged seven cents per pound of green leaf. Farmers were now actively taking up their opportunity to purchase shares in the tea factory companies in order to benefit from the dividend payments. All the shares allocated to farmers at Ragati factory were taken up by 1970, which led the company to make a further issue; at Nyankoba factory 29 percent of the allocation had been purchased.[16]

KTDA now made arrangements to increase the rate of planting in 1970–71. More dramatically, the Authority concluded that it had the resources and capacity to conclude the Third Plan a year earlier, in 1972, by a massive increase in the rate of planting. There was ample VP planting material (the Authority sold 9 million VP cuttings to Tanzania in 1971) and farmers had become more expert in the technique and in establishing new bushes. A total of 11,159 acres were planted in 1970–71 and as much as 14,744 in the following year, which brought the total smallholder tea acreage up to 65,788 acres.[17] The plan was completed on a good note under favourable weather conditions, with manufactured output some 60 percent higher than the previous year, at 11.3 million kg of made tea. With its finances in good shape, KTDA was able to reduce the green leaf cess on growers from 16 to 15 cents per pound at the start of the final year, which offset the impact of lower prices to some extent.

As against the figure of 130,000 in the Brown Report, KTDA now recognised that there were up to 150,000 farmers within the 'tea line' ecological zone who could plant tea.[18] It was expected that the Third Plan would encompass 66,000 of them by bringing in another 23,000. In the event, the number of tea farmers had risen to almost 67,000 (including those in settlement schemes), of whom 29,000 were new growers, an increase of 76 percent. Whereas, at the start of the period, Central Province farmers accounted for over 40 percent of the total and farmers in Kisii for around 15 percent, by the end of the plan in 1972 the Central Province share was down to 37 percent and Kisii accounted for 25 percent.

There were six tea factories in operation at the start of the Third Plan and ten by the end of 1972, but with a further 28 factories still due to be built to process the crops of the first three plans when they reached maturity. KTDA's financial position improved during the Third Plan as a result of growing cess revenues and the spreading of its operating costs over larger output. The consequence was that its external financing requirement from the IDA and CDC to cover its operating deficits was much reduced. Of the $4.9 million made available by the IDA, only about 35 percent was expected to be drawn, while most of CDC's loan commitments for field development were undrawn and lapsed.[19] This favourable development was already apparent by September 1970, which led the Authority to begin working on a Fourth Plan, although the Third Plan was only in its second year of implementation. There was, however, an early benefit to Kenya from its underutilisation of World Bank funds, which addressed a major concern of the Authority. The condition of tea roads was a perpetual worry since, in the absence of tarred surfaces, the roads required a level of maintenance which went beyond the resources of the Ministry of Works. KTDA incurred heavy costs from purchasing four-wheel-drive vehicles and it was forced to restrict the capacity of its tea factories below their optimum, in order to reduce haulage distances. It was therefore a welcome development in 1972 when agreement was reached between the Kenya government and the IDA that part of the unspent loan facility could be devoted to a programme of road improvement in the tea areas.

CDC to the Rescue

Having resolved to complete the Third Plan a year early, in 1972, KTDA presented a draft Fourth Plan to its board in August 1971; this was to plant another 45,000 acres over the three years to 1974–75 without the need for any additional external borrowing. As we have seen, the borrowing requirement was defined by reference to financing KTDA's annual deficits; however, the growth of its income from cesses on the farmers, coupled with effective cost control, had far exceeded expectations, mainly due to greatly increased plucking yields. The result was that, by the completion of the Third Plan, KTDA only expected to draw $410,000 from its IDA facility of $2.1 million, and £100,000 from its CDC facility of £450,000. It was therefore assumed that there would be no difficulty in drawing on these unutilised credit lines for the Fourth Plan planting programme.[20] KTDA itself was only peripherally involved in meeting the capital costs of new tea factories and so the draft plan made no proposals in this regard, although it was noted that the manufacturing implications of expanding to over 100,000 acres of tea would be to increase the number of factories from ten in 1972 to an eventual 58. The draft plan had not been discussed with the government before its presentation and endorsement by the board, and it was then forwarded directly to the IDA and CDC, driven by the perceived need to maintain the momentum of the planting programme. This proved to be a mistake, since it overlooked the elaborate appraisal requirements of both the World Bank and CDC, while the narrow focus on the planting programme left major issues in the air. It would take two years to resolve them.

As it happened, a World Bank team was in Kenya at the time, reviewing the smallholder tea programme, although it only received a copy of the draft Fourth Plan after the board meeting and its despatch to Washington and London.[21] The so-called supervision mission had three main areas of concern over the Fourth Plan proposals. First, there was a general concern over KTDA's financial model, which was predicated on an annual reduction of one cent per pound in the green leaf revenue cess – the Authority's main source of revenue. In the absence of such cuts there was little justification for external finance for the tea planting programme. The World Bank team was inclined to the view that augmenting farmers' incomes in this way was not a priority, since they were already doing well. Much more important, in the team's view, was the need for KTDA to begin building up financial reserves against unforeseen developments. It was calculated that if the cess was held unchanged over the five years to 1978 there would be no need for external borrowing, whereas an annual reduction in the cess of one cent would lead to a deficit in KTDA's accounts of £917,000 and a consequential borrowing requirement.[22] The World Bank (and CDC) had a veto on cess reductions through their loan agreements, and for several months it opposed any reduction. This became an embarrassment, since the Minister for Agriculture had already made a public announcement of the cess reduction for 1973–74 following the August board meeting. KTDA argued its case strongly, supported by CDC, who noted that labour costs had risen and also that farmers, especially new growers, needed a stimulus to encourage further planting.[23] Washington conceded.

A more substantial criticism was that the draft plan did not address the large factory finance requirement that was already building up under the earlier plans. This was compounded by KTDA's proposal that it would take over the management responsibility for all unbuilt factories in a new factory division. As noted, ten tea factories were in operation by the end of 1972 out of an intended 16. The increasing harvest from maturing plantings (leaving aside the requirement generated by a further 45,000 acres of tea), indicated a need for an additional 22 factories that would need to be built during the Fourth Plan, as well as 15 more later on.[24] Formally, the Kenya Government was responsible for arranging finance for this large capital investment programme, and there was an assumption that CDC would undertake a large part of it, stemming from its original commitment to the IDA in 1968 to provide half of the finance for 25 new factories (subject to several conditions). The supervision mission was concerned to discover that the Kenya Government and KTDA were thinking of excluding CDC from further factory financing, with a view to meeting its requirements exclusively from the IDA instead. Furthermore, the team's soundings with the major tea companies indicated little appetite for large-scale investment in KTDA factories.[25] It was also critical of the proposed factory division and doubted that the implementation issues had been fully addressed.

The third concern of the supervision mission was over tea prices and the world market prospects. The main tea-producing countries had been meeting for several years under the auspices of the FAO to consider the global supply-and-demand outlook, which was characterised by weak demand in the main export markets, especially the UK, and rising supply. The established producers were keen to restrict new planting and Kenya found itself under strong pressure, which it resisted. A major motivating influence in the hurried preparation of the Fourth Plan had been a concern that mandatory restrictions on new tea planting might come into force in the near future.[26] The World Bank was bound to be influenced by the FAO's assessment of the situation.

Notwithstanding these concerns, the supervision mission endorsed a request early in 1972 from the Kenya Government for the assistance of the World Bank's Nairobi office in preparing a full project submission, with a view to an appraisal mission coming out before the end of the year. It was quickly apparent that the delay would prevent KTDA including the 1972–73 planting season under the Fourth Plan. The solution was to request the IDA's and CDC's approval for their existing loan facilities to be available to support an 'extension' to the Third Plan to plant 15,000 acres of tea. This was agreed, although in the event only 8,500 acres were planted.[27] The delay in securing support for the Fourth Plan had a further consequence in respect of new factory finance: there was a need to build three more before the end of 1973. It had been assumed that their cost would be covered by World Bank and CDC loans, but it became clear during the appraisal mission that World Bank procedures precluded what would become retrospective financing.[28] The Kenyan Treasury was supposed to be the backstop, but no provision had been made for this eventuality. An appeal was made to CDC, supported by the World Bank, to meet the full external costs of the three factories, at a cost of £921,000. In this manner, a continuation of the tea planting programme in the gap year was finessed by its incorporation within the loan facilities already in place, and CDC also agreed to finance three new factories.

The KTDA board meeting on 10 August 1972 to consider the

revised Fourth Plan was an extraordinary occasion. The document had been prepared by the Ministry of Agriculture, with the assistance of the Nairobi office of the World Bank, as noted. It was only seen by the chairman and general manager of KTDA the day before the board meeting, and likewise by Peter Meinertzhagen, CDC's regional controller. There was no indication that CDC should participate in the financing of the plan, either in the planting of 45,000 acres of tea, or in the construction of 22 new factories. Reporting on the meeting, Meinertzhagen wrote: "Right at the outset of the discussion on the Plan, there was a minor uproar at the Board when Chairman Kimau and GM Karanja accused the Ministry of Agriculture of lack of cooperation with KTDA and going completely against the wishes of KTDA Board, which was to invite CDC to continue and strengthen its ties with KTDA."[29] The board proceeded to pass a resolution requesting CDC's inclusion in Fourth Plan financing and its participation in the forthcoming appraisal mission. Meinertzhagen surmised that the explanation for wishing to cut out CDC from a role in financing the Fourth Plan was that the Kenya Treasury wished to take all the external funding from the IDA at its 0.75 percent lending rate, and then benefit from the resulting spread of 6.5 percent when on-lending the money to KTDA. (CDC would be lending direct to KTDA and to factory companies under a government guarantee.) But it also emerged later that the World Bank was not averse to the government benefiting from this arbitrage, on the dubious argument that it encouraged capital formation.

In the weeks following the board meeting it was clarified that CDC should participate in the appraisal mission in November, with a view to participating in the financing of the Fourth Plan, although there was still considerable uncertainty within CDC as to whether the World Bank would make room for it.[30] CDC's aim was to provide 25–30 percent of the loan funds in respect of the 22 factories to be built during the four years to 1978. The appraisal mission duly went to Kenya in November 1972, by which time it had been accepted that KTDA could continue to reduce the green leaf cess by one cent per pound in each plan year. The World Bank and CDC had also accepted that KTDA should set up a factory division to manage the new factories; this was also a victory for KTDA over the Kenya Government, which had preferred an arrangement whereby the new factories would be built and managed by a separate corporate entity. The matter had to be referred to the president for a final decision, and President Kenyatta ruled in favour of KTDA.[31]

The Fourth Plan proposed to plant 36,500 acres of tea over four years to 1978, which would result in the total acreage of smallholder tea rising to 110,000. It was proposed to allocate 19,200 acres to 24,000 new tea farmers, so that there would be some 103,000 tea farmers at the end of the plan period.[32] In order to assist subsistence farmers who would be growing tea for the first time, it was proposed to provide the VP cuttings on credit to them (subject to a 15 percent cash payment), with repayment coming from a cess levied after the tea came into bearing in the fourth year. However, the most striking feature of the plan was the recognition of the need to provide a large number of new factories to process the growing volumes of green leaf now being harvested from the maturing earlier planting programmes. It was intended to have 16 factories (including Ragati) by the end of the Third Plan, although only ten were commissioned by the end of 1972. The Fourth Plan now proposed that 22 factories be constructed in the plan period to June 1978, leaving some 20 more for later on.[33]

As assessed by the appraisal mission, the total cost of the Fourth Plan would amount to K11.36 million, of which the external financing element was estimated at £7.9 million, which was to be provided as £4.8 million by the IDA and £3.1 million by CDC. This was duly approved by the CDC board, but within weeks the plan was in disarray. Washington telephoned to say that, under pressure from India and Sri Lanka and reinforced by pessimistic FAO forecasts of future tea prices, the World Bank board had refused to finance any new tea planting in Kenya.[34] However, the Bank was still prepared to contribute finance for the factories to be built to process tea planted under the earlier Second and Third Plans – 17 of them – and also to meet the costs relating to setting up the new factory division. CDC received a visit from Kenya government representatives with a plea to salvage the plan by taking over the field development element that had been rejected by IBRD, as well as five new factories relating to these plantings. There was a further blow: the loans would now come from the World Bank and not its IDA affiliate, at a cost of 7.25 percent instead of 0.75 percent; however, CDC loans would be made direct to KTDA instead of to the government and would carry a rate of 5 percent. CDC agreed to the rescue proposals and to increase its financial commitment to the plan from K£3.1 million to K£5.5 million (£6.2 million).[35]

In effect, there were now two programmes to be financed: one for the 17 factories required to process tea grown under the Second and Third Plans; and the other for the Fourth Plan field development and five more factories.[36] The factory project had an estimated cost of K£8.1 million and was to be funded 46 percent by the World Bank, 28 percent by CDC and 26 percent by the Kenya Government. The second programme for field development was costed at £4.4 million and was to be funded 75 percent by CDC and 25 percent by the Kenya Government. These proposals were formalised in Washington in May 1974. In the first three plans the World Bank had only been involved in the planting programme, through the IDA, and CDC had taken the main burden of arranging finance for the ten factories commissioned in the period. In the Fourth Plan, IBRD were now substantially engaged in the factory programme for the first time. CDC, for its part, became concerned that the new factory division, which only became operative in mid-1974, had not fully grasped the magnitude of the challenge of planning and building 22 factories, and in particular the timescale involved. Swynnerton drew up a planning schedule which demonstrated that some 26 months would elapse between site identification, placing contracts, construction and commissioning.[37] It was already clear by early 1975 that the factory programme was well behind schedule and that none of the four intended factories for that year would be completed in time. In the event, only nine factories were completed by 1997; in contrast, the field planting target was achieved in four years instead of five.

The factory division not only had the challenge of a large construction programme, but also the task of managing the new factory companies as they came on-stream. The ten factories built during the Second and Third Plans were managed by established tea companies under term contracts. In 1975 KTDA discovered that Brooke Bond Liebig had manipulated the accounts of its two managed factories to its own advantage, first by failing to pass on the benefit of freight rebates that it had negotiated, and second that by paying its own Mombasa office an unauthorised 1 percent commission on sales. With some reluctance the money was refunded, but this led to KTDA terminating their managing agency in 1976 and it resolved to take back all the managing agencies at the earliest opportunity.[38] At the end of 1977 there were still four factories with independent management contracts outstanding and KTDA was managing 16 factories.[39]

On the smallholder front a number of concerns had surfaced. The high price of fertilisers after the oil crisis had the consequence that farmers had virtually ceased to purchase them. KTDA's fertiliser supply scheme was suspended between 1973 and 1975 and harvest yields were suffering. In 1976 this led CDC to propose a scheme for a £600,000 revolving loan fund, whereby a farmer could repay his fertiliser purchase over 18 months via a cess. Its appraisal demonstrated that farmers would gain financially after borrowing costs through increased yields. There was a good – 60 percent – take up from established farmers.[40] There were still familiar concerns over tea roads, but also indications of variable standards in the advisory service to farmers. A field inspection visit in 1976 to 21 farms revealed average yields of 4,400 lb per green leaf per acre, ranging from 3,800 lb in Kericho to 5,000 lb in Embu, 5,300 lb in Sotik and 5,600 lb in the Nyeri sample.[41]

The Fourth Plan tea planting programme started off with a credit of 3,000 acres that had been planted in 1973 and this served to mask a distinct falling off in demand for new planting from farmers. The World Bank and CDC both came to the conclusion that farmers needed to have a larger regular income from the initial payments for green leaf, at the expense of the lump sum second payment after the close of the financial year.[42] The gross first payment of 90 cents per kilogram was itself subject to an immediate deduction at source for the KTDA cess to finance field development, which reduced the payment to 57 cents per kilogram. If farmers employed labour for plucking – and many did – the going rate for many years of 20 cents per kilogram was edging up towards 25 cents. And then there was the cost of fertiliser. In the mid-1970s the world market prices were high and, with ex-factory production costs of about Shs9 per kilogram of made tea, the factory companies were making sizeable second payments at the close of the season. In the three years 1973–76 second payments averaged 52.25 cents per kilogram, compared with an average of 14.40 cents per kilogram in the previous five years.[43] The World Bank and CDC pressed for a rebalancing of the payments, but this was resisted by the KTDA management for several years, on grounds of caution; however, an increase in the first payment to 100 cents per kilogram was eventually introduced from July 1977.[44] This was too late to influence planting demand for the Fourth Plan, which fell below expectations, notwithstanding the high world price of tea; nevertheless, the target of 25,000 acres was achieved, thanks to the carry-over from the Third Plan.

As well as setting up a factory division in 1974, KTDA also established a marketing department in 1975 to promote sales from its own managed factories – which at that time still only amounted to four. However, in the following year, four new factories were commissioned and the department then had the output of the two factories taken back from Brooke Bond Liebig in July, thus making ten in all. An experienced marketing officer was recruited and an international sales campaign immediately commenced, with visits to the UK, Europe, North America, Japan and the Middle East. The general manager went on all the visits. It was notable that one-third of sales were by private treaty and there was a similar proportion of private treaty sales locally, with the consequence that only 20 percent was sold through the Mombasa auctions. Although private treaty sales had the potential to command better prices, the policy was controversial. Moreover, when regard was paid to the total output of KTDA coming on-stream shortly, it was evident that more priority should be given to the Mombasa auctions. CDC's regional controller concluded: "My own belief is that over a period of time attempts to beat the market trend by private treaty sales are futile. It will be found that sales through the auctions are straight-forward and avoid the suspicion of under-hand arrangements."[45] The marketing officer was replaced and a policy was established that the bulk of sales should be through the Mombasa and London auctions, and that there should be no private treaty sales in Kenya.

The high prices obtained for tea exports in the early 1970s compared with the price set by Brooke Bond Liebig for Pool sales in the domestic market became a source of growing discontent in the industry, shared by both the estates and KTDA. There was a perceived conflict of interest for the Pool manager in that the company was paid for its Pool management role, even though its estates were on the same footing as the rest of the industry. Moreover, the price disparity was leading to tea smuggling to neigh-bouring Tanzania and Uganda, with resultant shortages of tea in the home market.[46] KTDA's relations with Brooke Bond Liebig had been soured by the scandal at Chinga and Ikumbi factories – and especially by the way it was initially handled by the company. It was no surprise that, early in 1977, the estate companies, together with KTDA, gave notice to the Pool that they would withdraw from it unless there was a price uplift. Within months an order was issued from the Ministry of Agriculture to KTDA to take over the domestic marketing of tea in Kenya with effect from the beginning of 1978. CDC suspected that this had been planned for some time.[47] It was decided to set up a new tea distribution company, the Kenya Tea Packers Association, or Ketepa as it was branded, which would be 60 percent owned by KTDA, with the balance owned by the estate companies. With an aggrieved counter-party, this was a rushed timetable and for some months it seemed as if Ketepa would have to build a new packing factory, for which it acquired a site; however, common sense prevailed and the Brooke Bond Liebig packing factory was sold to Ketepa.[48] In the event, KTDA only managed Ketepa and the Pool for one year, under stressful conditions. The accession of Arap Moi as president on the death of Jomo Kenyatta in 1978 triggered a sweeping patronage revolution in the civil service and with parastatal bodies. In this uncertain environment KTDA came under strong public criticism over tea shortages in the domestic market, including allegations of being directly involved in tea smuggling. Against this background the government decided that, with effect from April 1979, Ketepa should be an independently managed enterprise and not under the control of KTDA.[49]

Coping with Success

Already in 1977 CDC became aware that a Fifth Plan was in contemplation, to run until 1982, and which would add a further 25,000 acres of tea in existing tea areas and 10,000 new farmers.[50] The joint World Bank/CDC mission visit in November found the programme already under way and recorded that it would have preferred to see instead a focus on improving farm yields; however, it was somewhat mollified by an assurance that the plan was likely to represent the culmination of the expansion programme by the Authority.[51] Before undertaking this new planting programme, KTDA decided to carry out a tea bush census across its whole demesne. It had been aware for some time that there was a disparity between the licensed acreage approved in successive development plans and the actual planted tea acreage, as defined by the standard of approximately 3,500 bushes per acre, and that this was part of the explanation for lower than anticipated yields of plucked leaf per acre. The problem had occurred due to non-replacement of plants that had failed to establish themselves, and also because some farmers had failed to plant their licensed allocations. In 1981 the licensed tea acreage was 139,000 acres, whereas the census revealed that only 105,000 acres of tea had actually been planted – a shortfall of 34,000 acres, amounting to a quarter of the planting that had been licensed.[52] This was almost equivalent to the acreage that had been planted under the Third Plan and KTDA decided to implement an infill programme, spread over eight years to 1990, and thus to align the licensed and actual totals to an overall figure of 138,000 acres. This would also involve increasing the number of tea farmers from 143,000 to 150,000.

One consequence was that the Authority still needed to maintain a large agricultural department of field officers in the 1980s. The ratio of farmers to field staff was increased from 170:1 to 230:1, which implied a field staff of 650 – all seconded from the Department of Agriculture, but paid for by KTDA. An additional reason for maintaining this cadre of extension staff was the challenge of improving cultivation practice towards the standards achieved by the estate companies – i.e. the correct application of fertiliser and weeding, plucking and pruning routines – in order to improve yields and income. In general, standards were higher in the original areas in Central Province, as against the newer areas west of the Rift Valley. Overall, KTDA yields were around one-third of those being achieved by the estates, which represented a substantial loss of potential income to farmers. The government's inability to maintain tea roads to a serviceable standard was a recurring headache for KTDA, as it imposed heavy transport costs on the Authority. It was also a very visible demonstration of the lack of concern for rural betterment.

Although the Fifth Plan was implemented without a corresponding new factory element (and hence a need for external loan finance), there was clearly an implication that additional manufacturing capacity would be required in due course, both from the plan and from the infill programme. More immediately, there were concerns over the implementation of the existing new factory programme, and these were aggravated when the chief development engineer was removed in mid-1979 and not replaced, so that the department was under the direct control of the general manager.[53] The construction programme was well behind schedule and, in addition, cost escalation meant that there would be a financing shortfall for seven of the 22 new factories. KTDA had resisted the suggestion that priority be given to factory extensions on the not unreasonable grounds that this would apply to Fifth Plan planting when mature, whereas the new factory programme related to processing green leaf from earlier plans.[54] An application was made to CDC to finance four of the shortfall factories with an enhanced capacity of 1.2 million kg, which was approved, and then, six months later, CDC's regional controller concluded that further urgent action was required to increase manufacturing capacity of the five factories currently being built by adding a third drier, to raise it to 1.8 million kg, which was also implemented.[55] Even so, a World Bank mission in mid-1979 remained concerned by the looming pressure on manufacturing capacity in the coming years and that factories would be forced to operate well in excess of their rated capacity.[56] In the following year, a CDC mission envisaged that up to 20 more factories might well be required by 1990, while echoing concerns over the under-resourcing of the factory department and the backlog of planned maintenance.[57]

Side by side with the factory development programme, KTDA was also expanding its management role in respect of the operation of the tea factories. By 1981 it was responsible for managing 26 factories and, over the following years, its portfolio increased to 39 factories as the new ones were commissioned and the last remaining independent contracts came to an end. However, as foreseen by the visiting missions from the World Bank and CDC, the completion of the new factory programme was but the end of a phase and not the final requirement. The growing volume of harvested leaf threatened a capacity crisis that was only postponed by low yields and dry weather, and it was clear that a fresh programme would be needed. An appraisal mission was mounted in early 1987, the outcome of which was to identify an urgent need for four new factories with a capacity of 2.2 million kg by 1990, and a future requirement for up to 11 more to deal with the increasing crop, which was expected to increase by 40 percent by 1991.[58] Owing to the slow procedures of the World Bank it was proposed that the first four factories of phase one should be financed by CDC and the European Investment Bank (EIB) jointly and that the World Bank would participate in the later phases and approval was given on this basis. There followed a repeat of the 1979 episode when the board of the World Bank refused to sanction further investment in Kenya tea, with the result that CDC and the EIB were left to step into the breech.[59] Ironically, this did not prevent the World Bank being retained by the two remaining lenders as consultants on the factory programme. It recommended that, instead of building 11 more factories, a better solution would be to expand the capacity of 21 existing factories and to only build two more new ones.

These developments in the 1980s took place in an environment in which KTDA found itself increasingly in the public eye, which was unanticipated by management; it was also having to address a financial crisis in its own affairs for which it was also not well prepared.

Charles Karanja, the experienced and independent-minded general manager of KTDA, felt the full force of President Moi's desire for compliant institutions and loyal placemen when he was summarily removed at the president's behest in March 1981.[60] There had been indications for some time that KTDA's self-confident management had become less willing to listen to advice and was determined to demonstrate its autonomy, which made it vulnerable to the controlling regime in State House. Numerous decisions that were within the remit of the board were thereafter subject to clearance, and interference, from government.[61] In 1986 the State Corporations Act was passed, which formalised the dependent status of institutions such as KTDA and caused considerable concern to the World Bank and other external lenders over such provisions as these: all budgets and borrowing were subject to Treasury approval; financial surpluses could be appropriated by the Treasury; pay scales aligned to the civil service were imposed; and the composition of the board could be imposed.[62] The Regional Controller noted "Unfortunately KTDA is seeing increasing evidence of Government's wish to strengthen its control over it as a parastatal organisation" and the Office of the President had taken a place on its board.[63]

A further indication of the diminished status of KTDA came from two government initiatives concerning smallholder tea that were decided without consultation with the Authority, yet nevertheless involved it in their implementation. In early 1982 it was informed that "a decision has been taken to vest development of tea in Olenguruone under KTDA as with the other smallholder development zones."[64] A project had been prepared within the

Ministry of Agriculture and KTDA was told to identify financial resources for the project, including a tea factory. It emerged that the Agriculture Department had already established a tea nursery two years earlier and that the project involved a nucleus estate of 1,000 acres of tea and a settlement scheme to encompass another 3,000 acres on some 3,500 new farms. Olenguruone in western Kenya was quite remote and had no all-weather road access. It was estimated that some £2.6 million would be required to establish the project over six years, excluding the cost of road improvement. KTDA very reluctantly took on the assignment, while making clear that it had no finance to contribute and that it would have to be fully funded by the government. CDC noted, "It is understood that very influential people in Kenya have a stake in the scheme."[65]

A much more prominent indication of KTDA's dependent status occurred early in 1985 when President Moi announced that the forest zone protecting Mount Kenya, the Aberdares and Mau forest was to be cut back by 100 metres and planted with an estimated 18,000 acres of tea grown by smallholder farmers. It appeared that KTDA would be expected to establish tea nurseries, build the required tea factories and be responsible for tea marketing. Much of the land in question was likely to be marginal compared with existing areas on account of altitude and steep terrain.[66] The board was concerned to learn that its executives had already been participating in the preparation of the project and had been instructed to prepare the feasibility study and develop tea nurseries for planting to commence in 1986. In the event, KTDA's own nursery at Kangaita was effectively appropriated for the project and its officers were diverted to work on it.[67] Despite government agreeing to reimburse expenditure incurred by KTDA, there were lengthy delays in doing so. The Nyayo Tea Zone Development Corporation, as it was called, was eventually established in November 1986, reporting to the Office of the President. Its board of directors did not include representation from KTDA. As harvesting developed, KTDA factories were expected to take the green leaf, pending construction of new factories. A further concerning development occurred in 1988 when the chairman of the Nyayo Corporation announced that it would extend its operations ten to 15 kilometres into KTDA's scheduled areas and would recruit new farmers and also existing tea farmers. "The political delicacy of the matter was recognised."[68]

The management and board of KTDA could not expect to remain immune from interference in the political environment of the time, but they were slow to muster support when it mattered. This was dramatically illustrated at the end of 1988 in connection with the second payment to farmers that year. Tea prices remained depressed and far below the levels of the mid-1980s so that the amount available for distribution was about the same as in 1987 at Shs2.69/kg. Nevertheless, the board decided to impose a double deduction from this figure: 20 cents to meet the cost of its new headquarters and central warehouse and a further 16 cents to offset the operating loss in its field services, making a net average payment of Shs2.33/kg. This average figure masked a wide disparity between the results of individual factories, and especially the contrast between the high-yielding factories in Central Province and the much less profitable ones west of the Rift Valley, from where President Moi's main political support was drawn. There was uproar in parliament after the reduced payment was announced in late November and the Minister for Agriculture responded by instructing KTDA to reverse the 16 cent deduction to meet the field services deficit. He also announced that there would be an enquiry into the structure and operations of KTDA. At a subsequent board meeting attended by the permanent secretary of the Ministry he was outspoken and "attributed recent adverse comments on KTDA to poor public relations and he blamed management for doing very little to improve the image of KTDA in the eyes of politicians, the parent ministry and provincial administration... Management has been badly shaken and morale at HQ and in the field is low."[69]

In October 1989, President Moi announced the appointment of a committee to make recommendations on reorganising KTDA, with wide-ranging terms of reference.[70] When CDC was called to give evidence to the committee the main focus of interest turned out to be the mechanism whereby farmers could become shareholders in the factory companies. Richard Kemoli explained that its shareholdings were held in trust for the farmers and that it had been a willing seller for some time:

"Whilst welcoming the review I put across the point that fundamental restructuring of KTDA is perhaps not necessary. What is required is streamlining the existing structure so that a more efficient and responsible institution can emerge from the review. The Chairman appeared to take this view on board."[71]

The preoccupation with the delayed factory completions and inadequate processing capacity of the late 1970s abated in 1979–80 due to the effect of drought conditions on yields. Instead, the focus of concern became KTDA's own financial health. In 1980 its financial condition suddenly became precarious as a result of several adverse developments and was a source of serious worry for several years. These problems exposed unforeseen weaknesses in the way that the Authority had been structured, which affected its operating budget, its relations with the factory companies and its cash flow. Together, they threatened insolvency. The Authority was not itself a producing or trading enterprise; its role was essentially an enabling one, with agency functions. Thus, by 1980, KTDA had organised the introduction of a new cash crop to some 130,000 licensed farmers across the Kenya highlands, and was providing field advisory services and managing leaf harvesting collection centres; it had arranged the construction of 39 tea factories with borrowed capital, set up as separately managed factory companies in which it held a major shareholding that would eventually be sold to the farmers; and it had set up a marketing system for the manu-factured tea produced by the factory companies. Its main source of revenue to perform these varied operations was derived from the cess on farmers' green leaf harvest and was therefore solely a function of the physical size of the crop. However, this revenue stream had to offset expenditure that was affected by inflation and currency devaluation, namely the salaries of field staff and fuel and transport costs, which were on a rising trend. Although the total crop was increasing with maturing acreages, the drought of 1980 checked revenue growth and there was an overall operating loss. With the imminent completion of the Fifth Plan, there was the prospect that the cadre of over 600 field staff could be reduced to reduce costs, but the acreage-census exercise and the resulting infill programme meant that there was still a need for a widespread field force for a number of years. Moreover, there was a demonstrable need to improve harvest yields through better cultivation in order to get closer to the results being achieved on commercial estates. Another aspect was that KTDA farmers were the only agricultural-ists in Kenya who had to pay directly for the salaries of the field extension staff, who were all seconded from the Department of Agriculture. An obvious measure to help KTDA would be to relieve it of this cost. The board resolved to terminate the arrangement in 1982, but implementation was protracted.[72]

KTDA encountered two major financial problems over the factory construction programme. The first was that the Treasury was failing to pass on its agreed contributions towards costs, as well as the proceeds of the loans raised from OPEC and the EIB for six factories, for which contracts had been let.[73] In February 1983 work on the factories had to be suspended and, in turn, OPEC and the EIB paused disbursements to the Kenya government. In order to sustain the construction work KTDA borrowed from the factory companies from the funds they held for the second payment, but this had to be reimbursed before the distribution date in October.[74] A more serious complication arose from the financial arrangements for passing on loan proceeds between KTDA and the factory companies, which was exposed by the construction delays. The factory companies were only required to service their borrowings on completion of construction, and they were also accorded a three-year grace period before commencing repayment of the principal. Meantime, KTDA had a fixed schedule of interest payments and capital repayments commencing in 1981–82 and 1982–83, for which it was now unfunded. A further difficulty was that the loans from the World Bank, CDC and the EIB were denominated in foreign currencies, so that KTDA was exposed to the full exchange risk as the Kenya shilling weakened against them. The obligations of the factory companies were expressed in local currency and so the cost fell directly onto KTDA's accounts. Losses of K£3-3.5 million in 1980 and 1981 ballooned to K£21.5-22 million in 1982 and 1983.

The three international lenders were sufficiently concerned about KTDA's problems to mount a special mission in July 1983.[75] It recommended that the Kenya Government should accept responsibility for the exchange risk on the foreign borrowing on the grounds that the government benefitted directly from the tea sector's prosperity and its large foreign exchange earnings. It also recommended that the government should convert its own loans to KTDA into equity. For its part, KTDA argued strongly that the lenders should reschedule the loan repayments in order to match them with the delayed factory completions and their repayment grace period. The government did not accept these recommendations, while the three lenders for their part declined to consider any rescheduling. However, KTDA was permitted to implement three major reforms within its own sphere of competence: to raise its factory management fee from 3 to 8 percent of gross sales value of made tea; to raise the green leaf cess from 31 to 38 cents/kg; and to raise the rate of interest charged to the factory companies from 8.5 to 12 percent.[76] By 1984 the combined measures of the increased cess on green leaf and the higher charges to the factory companies had transformed the financial picture by means of an operating surplus of Shs98 million and the immediate crisis was over. For the first time, KTDA began to accumulate significant retained earnings.

The tea acreage achieved by 1990 under the Fifth Plan was close to target at 184,000 acres. New planting continued in the 1990s so that, by the turn of the century, KTDA smallholder sector embraced some 400,000 farmers with 225,000 acres of tea. By comparison, the estate sector at that time had some 87,000 acres of tea. KTDA was by far the dominant tea enterprise and was responsible for 60 percent of the national production of 240 million kilograms, produced by 45 constituent factory companies. The Special Crops Development Authority of 1960 had evolved over 40 years into a giant parastatal enterprise, yet it was still subject to government approvals (and political interference) with regard to appointments, salary levels and investment decisions. In 1992 the World Bank had signed an agreement with the Kenya Government on public enterprise reform that embraced five major parastatals, including KTDA, in order to liberalise and privatise them. With tea, it took seven years to carry this into effect in 1999, with the publication of a sessional paper and ensuing legislation.[77] The governance structure of the industry was revised through a reform of the Tea Board, and KTDA was privatised in 2000. The privatisation was achieved by arranging that KTDA was owned by the 45 factory companies, which in turn were owned by the smallholders supplying each factory. This was achieved by KTDA (and CDC) selling their shares in the companies to the farmers. The outcome was that KTDA (now named 'the Agency') became a companies-act enterprise. It entered into formal management agreements with each factory and became a classical management agency, as seen in India in the 19th century, providing a range of services to the factory companies: factory management with seconded staff; extension services, again with seconded senior staff, while junior staff became employees; marketing and sales. Once factory companies had repaid their construction loans they would be entitled to appoint different management agents, or none. At the time of privatisation KTDA was in the process of developing a large warehousing facility in Mombasa; it was the controlling shareholder in Ketepa; and it owned an insurance broking company.

It is beyond the scope of this chapter to narrate how KTDA has discharged its new managing agency mandate in the 21st century. However, in conclusion, perhaps it would be useful to note some of the issues it will face in future. As managing agent to the factory companies – initially 45 of them, but expanding to 65 in the next 15 years – the operating agreements enable it to control manufacturing standards. However, the individual factory boards have to agree to capital expenditure to keep abreast of technical advances; these boards will always be under pressure to maximise payments to the farmer/shareholders. Hence the importance of the agent/factory relationship within the framework of the governing agreements. It is not difficult to envisage situations, once the companies are free of loan guarantees, where some might turn to the estates sector in future as an alternative. The problem of improving farm yields and leaf quality is more intractable, given the huge number of farmers involved (now in excess of 500,000). Yields in the estate sector are on average twice those of the smallholders, while many estates do even better. This represents a huge loss of potential income to the smallholders.[78] KTDA is no longer able to maintain an army of field extension officers, nor are the individual factories, so other approaches have been explored.[79] Kenya is fortunate in having embedded a culture of fine plucking in the smallholder sector. It has also been a world leader in the introduction of money transactions via mobile phones, which has enabled KTDA to introduce IT systems that deliver green leaf payments directly to farmers.

As the controlling shareholder in Ketepa there is a persuasive case to move to outright ownership, and to rationalise the respective responsibilities of KTDA and Ketepa for promotion and sales.[80] It is a commonplace to stress the need for Kenya teas to establish a stronger identity in export markets, and for there to be more vigorous export market development. Realising these laudable aims is less susceptible to easy answers.

During the Moi era and beyond, Kenyan public life became a byword for corruption and the looting of public institutions. KTDA was largely unscathed in this respect, but nevertheless it has had its quota of scandals arising from venal executives exploiting its purchasing of fertilisers and other farm and factory inputs.[81] Having been a government-controlled enterprise for many years, and hence subject to parliamentary questioning, Kenya's MPs have continued the practice of raising issues about KTDA in parliament. KTDA's very prominence also makes it a natural target for media interest and public lobbying by interested parties, including those with a political agenda. This is a distraction for management.

Having such an enormous shareholder base would be a major challenge for any public company, the more so as the leaf payments to farmers represent their main cash income in most cases. Explaining the relationship between realised auction prices and the amounts received at the farm gate will always be a difficult task for management, and a topic easily subject to populist distortion, not to mention the cost of management overheads and factory costs. The gap between living standards on a small farm and those at the higher reaches of management is all too easily open to questioning. In short, the move to private sector company status has substituted one kind of harassment for another in the management of one of the outstanding institutions in Africa.

NOTES TO CHAPTER 11

[1] MP/Library/19, *The Operations and Development Plans of the Kenya Tea Development Authority*, Nairobi, 1964.

[2] MP/CDC/3, *Proposals for Third Development Plan 1968/73*, KTDA to Minister for Agriculture, 23 February 1965.

[3] MP/Library/18, *A Report on the Tea Growing Potential of Kenya*, LH Brown OBE, Nairobi, 1965.

[4] MP/CDC/3, KTDA Board paper 24/66, 15 March 1966, comprising *Proposals For Third Development Plan 1969-73* and a covering paper from Swynnerton. Also Swynnerton to Regional Controller, 4 April 1966.

[5] MP/CDC/3, *The Third Plan*, KTDA Board paper 46/66, 29 June 1966.

[6] MP/CDC/3, 'Kenya Tea Development Authority', CDC Board paper 72/67, 13 July 1967.

[7] MP/CDC/3, Noted by Swynnerton on 23 March 1967 after he had met World Bank officers.

[8] MP/CDC/3, Board paper BP 72/67.

[9] BP 72/67, ibid.

[10] Chapter 1.

[11] MP/CDC/4, Board paper BP 3/68.

[12] MP/CDC/4, *The Operations and Development Plans of the Kenya Tea Development Authority*, KTDA Nairobi 1964; CDC BP 72/67 and BP 3/68 .

[13] MP/CDC/4, *Report of the Working Party on the Tea Industry*, Ministry of Agriculture, 22 September 1969.

[14] *Working Party*, ibid.

[15] MP/CDC/4, IBRD/IDA Supervision Mission Covering memorandum and report, 9 December 1968.

[16] MP/CDC/5, KTDA Annual Report 1969/70, BP 42/70.

[17] MP/CDC/5, KTDA Annual Report 1971/72, BP 34/72. The Authority assumed responsibility for tea grown on the settlement schemes in 1971 and their figures were thereafter incorporated in the KTDA's. This added 3,719 acres to the total and another 1,759 farmers.

[18] MP/CDC/5, *Future Development of Smallholders Tea*, KTDA, 15 September 1970.

[19] MP/CDC/5, IBRD appraisal of the Third Kenya Tea Development Authority Project, March 1973.

[20] MP/CDC/5, *Fourth Plan Projections*, KTDA, BP 40/71, August 1971.

[21] MP/CDC/5, J Willems, head of mission, to DJ Parsons, head of agriculture, IDA, 14 September 1971; note from CDC office to London following meeting with Willems, 19 January 1972; CDC London to Nairobi also following a meeting with Willems, 2 February 1972.

[22] MP/CDC/5, Willems to Parsons, 6 March 1972; CDC EMB 181/72, 27 June 1972.

[23] MP/CDC/5, General manager, KTDA to permanent secretary, Treasury, 22 June 1972. CDC Nairobi added commentary notes on its copy of the letter.

[24] Reconciling planned factory needs, as dictated by the field development programme and the time taken for tea bushes to come into full leaf production, with the actual construction of new factories is not straightforward. Timing was further complicated by decisions to increase the capacity of existing factories, which had to be reconciled with the need to constrain leaf transport costs. The following tabulation illustrates the time lag between new factory require-ments of the planting programmes and their eventual commissioning.

	Plan factories	Built during plan	Cumulative	To come
Ragati factory			1	
1st Plan	6	–	–	5
2nd Plan 1963–64 to 1967–6	10	5	6	10
3rd Plan 1968–69 to 1971–72	–	4	10	6
3rd Plan extension 1972–73	3	–	10	9
4th Plan 1973–74 to 1977–78	19	10	20	18

[25] The CDC's regional manager was indignant, rejecting the suggestion that the government would consider excluding the CDC and asserting that his own contacts with the tea companies indicated that they wished to be offered further management and investment involvement in the factory programme. Subsequent events showed Meinertzhagen to be mistaken. MP/CDC/5 Regional Controller to London, 30 March 1972.

[26] MP/CDC/5, Willems, 6 March 1972.

[27] According to the KTDA's *General Review of the Year* (BP 36/73), 4,667 ha of tea were planted (11,500 acres), but the subsequent documentation relating to the Fourth Plan records that only 8,500 acres were planted in 1972–73 and this appears to be the more reliable figure. Cf. *Proposals for further Field and Factory Development*, EMB 103/73, 2 May 1973.

[28] MP/CDC/5, KTDA note on discussions with the appraisal mission, 11 December 1972.

[29] MP/CDC/5, Regional Controller to London, 11 August 1972.

[30] MP/CDC/5, Swynnerton to Regional Controller, 26 October 1972.

[31] MP/CDC/5, *Management of KTDA Tea Factories*, BP 20/72, May 1972.

[32] MP/CDC/5, *Proposals for Future Field & Factory Development*, EMB 103/73, 2 May 1973.

[33] MP/CDC/5, EMB minutes, 10 July 1973. Swynnerton to Regional Controller, 31 July 1973.

[34] Ibid.

[35] Including the three factories it had already agreed to finance in 1973.

[36] MP/CDC/6, Summarised in Swynnerton's brief for General Manager, January 1975.

[37] MP/CDC/6, Swynnerton to Regional Controller, 27 March 1975

[38] MP/CDC/6, Regional Controller to London, 3 April 1976. In the event, this took longer to achieve.

[39] MP/CDC/6, IBRD/CDC Joint Mission Report, November 1977.

[40] MP/CDC/6, *Proposed Loan for a Growers Fertilizer Credit Scheme*, EMB 54/76, 9 March 1976.

[41] MP/CDC/6, *Report on Field Visits and to Tea Factories*, CDC, March 1976.

[42] MP/CDC/6, IBRD Inspection Report July 1976. This represented a reversal of the previous view of the World Bank.

[43] MP/CDC/6, KTDA Annual Report BP 38/75 and IBRD Supervision Report May 1989.

[44] MP/CDC/6, Regional Controller to London, 14 March 1977. The KTDA Board approved an increase to 98 cents/kg as from July, which was subsequently rounded up to 100 cents/kg.

[45] MP/CDC/6, Regional Controller to London, 8 February 1977.

[46] The contrast was between a local retail price of Shs14.50/lb and a Mombasa auction price of Shs26/lb.

[47] MP/CDC/6, Regional Controller to London, 14 March 1977.

[48] MP/BB/3, Brooke Bond Liebig Kenya Annual Accounts 1978. The company realised a loss on cost of £1.4 million.

[49] MP/CDC/6, KTDA Board minutes, February 1979.

[50] MP/CDC/6, *The Proposed 5th Plan*, KTDA, BP41/77, December 1977. It was notable that the plan did not allow for any more factories.

[51] MP/CDC/6, Joint Mission Report, November 1977 ibid.

[52] MP/CDC/7, Note on KTDA board meeting, 17 April 1986.

[53] MP/CDC/6, EMB 145/80.

[54] MP/CDC/6, *Additional Finance*, BP 80/78, 9 November 1978.

[55] MP/CDC/6, *Additional Finance*, EMB 114/79, 1 May 1979.

[56] MP/CDC/6, IBRD Supervision Mission, May 1979.

[57] MP/CDC/6, *Review Mission of KTDA*, March 1980.

[58] MP/CDC/7, *Second Tea Factory Project*, Joint Mission aide memoire, July 1987.

[59] MP/CDC/7, Exco 175/89 31 May 1989

[60] MP/CDC/7. "Charles Karanja was unceremoniously dismissed as GM and one of his assistant managers, Simon Kamuyu, has been appointed by the President's Office to take his place." Regional Controller, 16 March 1981.

[61] MP/CDC/7. "The mission gained the impression from the General Manager that there is increasing interference in the running of KTDA by the Parastatal Committee (Office of the President). The mission gathered that since the departure of the previous General Manager, who was insistent on autonomy, the KTDA Board has lost some of its decisiveness and that its role has become more advisory." IBRD Supervision Mission to Washington, 15 July 1981.

[62] MP/CDC/7, IBRD/CDC/EIB Draft Mission Report, March 1987.

[63] MP/CDC/7, Regional Controller to CDC London, 12 February 1987.

[64] MP/CDC/7, Director of agriculture to KTDA, 2 February 1982.

[65] MP/CDC/7, File note copied to London, 6 July 1982.

[66] MP/CDC/7, Minute of KTDA board meeting on 18 April 1985 and Regional Controller's note to London, 19 April 1985.

[67] MP/CDC/7, Regional Controller to CDC London following board meeting, 15 August 1985.

[68] MP/CDC/7, Regional Controller's file note following KTDA Board meeting, 30 September 1988.

[69] MP/CDC/7, Note of KTDA Board meeting on 5 December to CDC London, 7 December 1988. Also press cuttings.

[70] MP/CDC/7, File note forwarded to CDC London, 18 October 1989.

[71] MP/CDC/7, File note forwarded to CDC London, 27 October 1989.

[72] MP/CDC/7, Note on KTDA Board meeting, 1 July 1982 The Authority stopped paying the Ministry of Agriculture some years before the matter was resolved. IBRD Supervision Report, Annex 7, 26 November 1984.

[73] MP/CDC/7. In August 1982 the amount held back by the Treasury amounted to K£1.36 million; in November 1982 it was K£1.5 million and in February 1983 K£2 million. CDC reports to London, 27 August 1982, 24 November 1982 and 22 February 1983.

[74] MP/CDC/7, Note to London on KTDA board meeting, 22 February 1983.

[75] MP/CDC/7, IBRD letter to Ministry of Agriculture, 13 June and CDC note on the proposals, 17 June 1983.

[76] MP/CDC/7, Regional Controller to CDC London, 24 November 1983.

[77] *Liberalisation and Restructuring of the Tea Industry*, Sessional Paper No 2 of 1999.

[78] As a reference figure, total payments to smallholders in 2000 amounted to Shs24.6 billion, equivalent say to £184 million, so that a 10 percent improvement in average yields would put £18 million of additional purchasing power into the rural areas, other things being equal.

[79] A recent innovation has been organising Farmer Field Schools, led by leading farmers, to pass on improved practice. Another has been the establishment of farmer committees at the more than 3,000 green leaf buying centres to control quality and delivery to factories, as providing the basis of the governance pyramid.

[80] At its inception, the KTDA was a 60 percent shareholder in KETEPA; by 2010 this had risen to 80 percent.

[81] There was an incident in 1990 when the KTDA received an order from the Treasury to transfer to it the funds that had been accumulated for the second payment to growers – some £19 million. The Board met, refused to comply and got away with its lese-majesty. MP/IM/1, Interview Gatabaki, 17 February 2004.

12

THE IMPACT OF
INNOVATION AND SCIENCE

During the course of the 19th century a system of tea cultivation was perfected in India on large estates with their attendant factories that mimicked the procedures of Chinese hand-made tea. Therefore, by the time that Brooke Bond and James Finlay came to develop their new properties in Kenya, there was almost 100 years of experience to draw upon, and tea growing might reasonably have been regarded as a mature industry.

For twenty-five years the cultural practices of India and Ceylon were applied fairly consistently to East Africa. Tea plants were grown in tea nurseries from seed originally supplied from India. Two-year-old seedlings were transferred to large fields and planted in rows either four or five foot apart, with five foot between rows in the deep volcanic soil underlying the cleared rain forest. The young bushes were shaped as they grew to produce a flat table and leaf canopy some three foot high. Every four years the bushes were ruthlessly pruned by hand down to two foot and crop was lost for that year while growth resumed. The fields had to be weeded regularly until the canopy was established, especially to control the dreaded couch grass that flourished in Kericho which, if not cleared, would compete with the tea bush for water and nutrients. During the Second World War there were severe labour shortages on the estates and the couch grass invasion became a serious menace. A measure of the effort involved in digging out the grass around the roots to a depth of several feet was that the daily task per man was only five bushes; meanwhile, the tea bush was damaged and left with perhaps only three of four main roots.[1]

By the 1940s the cultivation practices and growing conditions in Kericho could generally be relied on to produce 1,000 kg of made tea per hectare, except that every four or five years the tea bushes had to be drastically pruned by hand in order to re-establish a low plucking table. This meant a loss of output for a year and also an opportunity for couch grass and other weeds to return. Following best practice in India, the estates were progressively planted with shade trees, since it was thought this simulation of tea's natural habitat was beneficial. Kenya tea proved to be mercifully short of pests and disease, apart from a potentially troublesome root fungus, *armillaria*, if the original forest tree roots had not been properly cleared away. Weather shocks could have a serious economic impact in that Kericho is subject to regular and violent hailstorms which strip the leaf canopy, and it can take a fortnight or more for an estate to recover. Periodic drought was the other worry, since under such conditions the tea bushes would cease to flush regularly and yields would fall. An occasional frost at Kericho's altitude of 6,000–7,000 feet could also 'burn' the tea flush.

The lengthy history of incremental improvements in field practices and in factory machinery accelerated dramatically in the 1950s as a result of research advances. At the same time, there was a growing awareness of the need to raise productivity in a labour-intensive industry, which was seeking to apply horticultural standards to large-scale cultivation. The new understanding of the physiology of the tea bush and of the chemical changes that take place when leaf is harvested and turned into a beverage have progressively been translated into operational practice.

The first major breakthrough in the 1950s was the application of effective herbicides that would eliminate couch grass and other weeds. By the 1960s, hand weeding had been largely eliminated as a field task. This development in turn cleared the way to soil enrichment: leaving pruning brash to rot and applying manure (especially in Limuru, where a cattle herd was maintained for many years), but more importantly the application of fertilisers containing nitrogen, phosphorus and potassium, and eventually other elements as well. Finding the optimum application – in both cultural and economic terms – has been the subject of many years of field trials, and is still being refined. Rather short-sightedly, the tea companies were jealous about their own field trials: instead of treating this as an industry-wide issue, they guarded information and were somewhat reluctant supporters of the Tea Board's Tea Research Foundation. By the end of the century each company had its own preferred fertiliser formulation, and James Finlay was applying it by aeroplane three times a year to save labour. A visit to the Tea Research Foundation in 2003 found its director critical of the design of the companies' field trials, as well as a view that their application dosage was too high, leading to excess surface root development at the expense of deeper roots. The soil chemist there was interested in the role of micronutrients and their possible link with microbial activity.[2] The companies thought this overly academic.

In the wild, *Camellia sinensis* is a 30-foot forest tree.[3] In cultivation it is subjected to close planting and savage pruning to establish a continuous leaf canopy, but there remains the woody frame and deep roots of a tree. Given that the economic objective is to maximise the leaf harvest, the cultivation regime needs to ensure that the growth energy from photosynthesis is directed to leaf production rather than root development. Such understanding led to a number of improvements in field practice. It made sense to increase planting density – thereby enlarging the leaf canopy – by inserting an extra row of bushes between the lines, so that the matrix was 5ft x 2.5ft or 4ft x 2ft. Attention was also given to pruning technique. The traditional drastic hand surgery of the bushes was both labour-intensive and vulnerable to a significant incidence of industrial injury, which provided an incentive to try mechanised pruning. It was eventually demonstrated that this did not adversely affect either the health of the tea bushes or crop yields, and it gave a significant productivity boost. The move to a less drastic pruning technique, and eventually to mechanical pruning, did less damage to the bushes and enabled faster recovery.

Research in India had demonstrated that the efficiency of leaf photosynthesis in tea was affected by light intensity, and that a large part of the product of photosynthesis was subsequently lost in photorespiration, rather than being mobilised by the leaves to support new growth. A practical consequence in Assam was that yields were improved by introducing shade trees to reduce photorespiration. This wisdom was then transferred to East Africa, but it was not until the 1960s that it was clearly demonstrated that this was an error. Conditions in Kenya were not as extreme as those in Assam and the effect of shading the tea was actually to reduce yields, apart also from an increased risk of root fungal infection. All the shade trees were then removed in Kericho. This was an interesting demonstration of the peril of transplanting a valid scientific finding from one set of conditions to a different environment.

Kenya needed to establish its own research capability and the Tea Research Institute was started in 1949 in Kericho by Brooke Bond and subsequently taken over by the Tea Board. However, both James Finlay and then Brooke Bond later set up their own research facilities and tended to guard their findings jealously. The Foundation then suffered from a lack of clear mandate and direction.

Tea quality had always been associated with selective leaf plucking. The Indian target of 'two leaves and a bud' was too refined for Kericho, which had long accepted four or five leaves and a plucking cycle of ten days (James Finlay) or 14 days (Brooke Bond).

The challenge now was how to bring about improvement in plucking standards and the necessary change in field operations. As we have seen, the traditional practice was that gangs of pluckers (originally women and juveniles) worked through the fields filling a basket on their backs with handfuls of tea leaves. They brought their loads to the weighing stations and were paid on a piecework basis, which itself was an incentive to coarse plucking. There was a sanction of rejecting unduly coarse loads, with a consequent reduction in the piecework payment, but this implied close supervision.[4] In Malaysia, the rubber plantations had been successful in fostering a link between the rubber tappers and particular trees: could this be replicated in Kericho? Brooke Bond ran a trial of fine plucking (two leaves and a bud) on an eight-day plucking cycle which was found to produce a higher leaf yield than a 14-day cycle and three or four leaves, but it required a higher labour input, which was a negative. And then James Finlay introduced a system whereby plucking teams became responsible for designated fields, with an individual being allocated a specific number of tea bushes to tend on a seven-day cycle. This was found to encourage a higher standard of plucking, since coarse plucking reduced the yield on the subsequent visit. It also encouraged family involvement in periods of heavy flush, or of illness. Brooke Bond then adopted a similar practice.[5] The regime of hand plucking of tea is now under challenge from rising labour costs and the advent of mechanical harvesting, so that the traditional picture of smiling tea pluckers may become as outdated as Millet's paintings of women sowing seed from their aprons and hand-held sickles for grain harvesting. In 2003 Unilever was carrying out mechanised trials on its properties in Tanzania – away from trade union sensitivity. Yet James Finlay was openly trialling rival techniques with worker cooperation. The most effective machine was a hand-held cutter pushed over the canopy and requiring three men; it had a notional yield of 1,200 kg/ha and required only 10 percent of the labour input of the traditional plucking teams.[6] Mechanical harvesting is already well established in some other tea-producing countries with a different tradition, such as Japan, Turkey and Argentina. It seems only a matter of time before economic pressures lead to its adoption by the Kenya estates sector, where wage rates are negotiated on an industry basis for a unionised workforce. The outcome has typically been an arbitration award that is heavily politicised.[7]

After establishing conditions for the application of fertiliser to boost leaf harvesting and recovery from pruning, the next great productivity boost came from the realisation that there were great genetic variations in different strains of *Camellia sinensis*: some plants were much more vigorous and others produced better-quality tea. Vegetative propagation from superior varieties had started in India and Sri Lanka in the early 1950s and James Finlay sent a manager to India in 1952 to study the new technique. Work commenced in Kericho, which led to the selection of seven varieties for mass propagation after rooting trials and manufacturing tests. By the mid-1960s all new planting was with plants grown from cuttings.[8] Brooke Bond was much slower in taking up variety selection and VP, despite an early start in south India, which seems to have been due to discouragement of the exchange of technical information between different parts of the group.[9] Nevertheless, Kericho engaged a scientist from India to select varieties, but he returned after a short while following a robbery incident, and the research programme lapsed.[10] The result was that Brooke Bond continued with mixed seedling planting for much longer. One informed estimate is that Brooke Bond ended up with only 25 percent of its tea comprising selected varieties, as compared with 40 percent at James Finlay.[11] The contrast with KTDA and the smallholder sector could not have been more striking: after initial planting on the traditional basis in the First and Second Plans, KTDA committed to using only selected varieties in 1965 for the Third Plan and thereafter. On a comparable basis, more than 80 percent of smallholder tea comprises selected varieties.

The selection of varieties for VP was based on intelligent observation and selection by estate managers, rather than on a systematic research project, but the outcome was that both tea companies identified several varieties for mass production with high-yield characteristics as compared with traditional mixed-seed fields. Combined with the other improvements in cultivation, the new varieties were capable of yielding between 5,000 and 7,000 kg/ha over a four-year cycle, and sometimes even better.[12] A case study from Brooke Bond of 30 years of comparative records of two fields illustrates the combined impact of improved cultural practices and the planting of a high-yield variety. Field A comprised seedling tea of mixed genetic parentage and illustrates purely the consequence of improved cultural practices. The yield from the field improved from 800 kg/ha in 1965 to a best of 4,500 kg/ha in

1996, a five-and-a-half-fold improvement. Field B was planted in 1970 with selected variety, which quickly yielded 2,500 kg/ha within three years – and the same as Field A in that year. But the yield went on improving to peak levels of 8,500 kg/ha in 1986, which was nearly twice the best that was achieved by the seedling tea, although it received the same cultural treatment. Meanwhile, some fields had achieved yields of up to 10,000 kg/ha.[13]

One characteristic of a tea bush that has been grown from a cutting is that it does not develop a tap root to reach deep into the soil. During two severe droughts in 1997 and 2002, and lesser ones in 1998 and 2004, the tea grown from cuttings suffered badly. With the prospect of global warming leading to more frequent droughts, there has been a heightened research focus on selecting drought-resistant varieties for future planting, but also a realisation of the need to establish from the outset a strong bush-and-root system. This entailed the elimination of fertiliser application to young plants and lighter pruning in the early years; it also led to reducing the planting density, as compared with the traditional 17,000 bushes per hectare.[14] There has been a growing interest in plant breeding to create varieties with planned characteristics, notwithstanding the lengthy timescale involved. This includes research to link chemical ingredients with flavour characteristics, which enables speedier variety selection. At the time of my visit, Brooke Bond had 48 new varieties coming up for release to the estates. As already noted, both Brooke Bond and James Finlay maintain their own research facilities in Kericho, notwithstanding the Tea Board's research institute there. The latter appeared to suffer from resource constraints and unclear priorities.[15]

Tea quality had long been associated with selective plucking: the 'two leaves and a bud' target. Improved understanding of plant physiology now demonstrated that the chemical compounds associated with quality decline with leaf maturity. The operational deduction was to have shorter plucking cycles, but how also to improve plucking standards? The traditional pattern of gangs of pluckers and young people working through the fields was not conducive, although rejection of coarse plucking at the weighing stations – with consequent reduction in piecework payments – was a partially effective sanction.

The new high-yielding tea varieties selected and then multiplied by VP were planted by the estates companies wherever they had the opportunity to increase their acreage. Replanting of old tea entailed a more difficult calculation, since the tea tree is a long-living plant and much of the original planting from the 1930s is still productive. There is a significant cost in grubbing up old tea, and loss of production for several years, all to set against the promise of higher yields. The financial calculation of clearing costs and waiting time versus lost production does not support replanting, yet there remains the long-term issue of regenerating the estates. The outcome has been a pragmatic decision to undertake affordable annual replanting programmes. Meanwhile, the delay has provided room for other research to prove itself, such as cultivation methods to establish more robust bushes and root systems, and the plant breeding already referred to.[16]

The result of some 40 years of innovation and applied research has been an agricultural revolution in the cultivation of tea. The apparent uniformity of the rolling expanses of vivid green tea fields that confronts the visitor is deceptive in that it is underpinned by a regime of horticultural intensity. Every field is treated individually: depending on the varieties of tea growing on it, the condition of the soil and its location, there is a specific fertiliser application for that field and the rows of tea bushes are tended and plucked by dedicated teams. Meticulous records are maintained. It is market gardening on an industrial scale. There is perhaps some irony in that this individ-ualised cultivation is mirrored, in concept at least, in the smallholder tea sector, where farming families tend their own plots of tea. Mainly, but not entirely on account of inadequate fertiliser application, yields in the smallholder sector are far below those achieved on the large estates, with a consequent shortfall in potential income. It is a major challenge for KTDA to close the gap between best practice on the estates and that in the smallholder sector.

The huge increase in the volume of harvested green leaf that resulted from the improvements in field management presented an acute challenge to the factories on the estates. For example, with James Finlay, the total crop increased from 4.9 million kg in 1964 to over 10 million kg by the end of the century. Using traditional manufacturing methods, this implied a huge investment in additional factory capacity. Could the traditional batch handling of the leaf harvest, involving manual transfer from one stage to another, be streamlined into something approximating to continuous throughput? And could this be achieved without sacrificing quality? As had always been the case with developments in tea manufacturing, improvements were incremental and were usually the product of ingenious factory engineers, which were then adopted by the machinery manufacturers. With this in mind, a number of innovations in the manufacture of tea fall into place.

The traditional method of withering the green leaf brought in from the field in order to reduce its moisture content was both labour-intensive and extravagant of space, since the harvested leaf was conveyed to the top of the factory building by hoists and laid out on racks, or tats as they were called, to take advantage of natural airflow. One innovation in the 1950s was to hang shelf units from an overhead rail which could then be rolled through a heated chamber, which was both space- and time-saving.

It had been discovered that, apart from the desired dehydration that occurred during withering, important chemical reactions were also taking place as the leaves used up their reserves of starch: the caffeine content increased but the theaflavins that were responsible for the brightness of the tea liquor and its flavour strength diminished the longer the process lasted. Thus there was merit in speeding up the dehydration period. This was achieved by the introduction of mobile troughs to hold the green leaf as it was moved along a ring main of runners, which also simplified the loading process. The troughs could then pass through a heated chamber or, better, ducted warm air could be blown over the troughs, so that there was then a process of continuous withering.

To initiate fermentation, the original rolling machines made of brass and then stainless steel replicated the action of two hands from the hand-made Chinese method of tea making, with single- and then double-action rollers. This was also a batch process until the innovation of the McTair rotorvane in 1958 and subsequent improvements. This was essentially a rotating barrel with vanes inside it, so that leaf could be fed into it at one end and emerge at the other. However, a more fundamental invention from Assam had been awaiting its day since the 1930s – known by its initials CTC, standing for cut, tear and curl – which achieved integration of the withering and rolling operations. Adapted originally from the tobacco industry (the Legg Cutter) and from wheat milling, the equipment comprised a pair of cylindrical rollers with grooves cut into them which rotated in opposite directions at different speeds. Initially the process was strongly resisted by tea brokers and tea tasters, especially for high-quality teas, but its production merits gradually prevailed. By the 1960s, CTC machinery had become essential for dealing with the increasing volumes of green leaf harvested on the Kenya estates and with KTDA. Brooke Bond installed its first plant in 1968 at Tigabi factory.[17]

The next challenge was to integrate fermentation into the processing sequence by feeding the wet leaf onto a moving bed, where an airflow could be forced over and through it in order to foster the oxidation reactions, instead of leaving it in a separate room. Here, control of timing was a key part of the art (and increasingly the science) of tea making. At one level of understanding it was known that short fermentation was associated with the liquor qualities of 'brightness' and 'briskness', while a long fermentation was associated with 'colour' and 'body', albeit with a duller liquor. At another level, the enzymatic oxidation of catechins into theaflavin (brightness) and thearubigen (colour and body) led to a changing balance as fermentation continued. The fermented leaf now had to have its remaining moisture removed in tea-drying chambers. The tea-drying process had been well mechanised since the 19th century, apart from hand loading the fermented leaf. Then, in the 1970s, there came an innovation from Sri Lanka, the so-called fluid-bed drier, in which the now broken up leaf particles are desiccated in a chamber of blown hot air. Two of these new machines could do the work of three old driers. The now dry tea was traditionally sorted into grades – pekoe, broken orange pekoe, fannings, dust – by mechanical sieves, but there was still a problem of removing excess stalk and fibre, which traditionally had been a manual operation. In principle it was known for some time that stalk could be removed by electrostatic attraction onto plastic rollers, but difficulties were encountered in making this work efficiently under factory conditions. The problem was solved serendipitously as a result of a machine malfunction, when the charged plastic rollers would only rotate at a very slow speed; under this condition it was discovered that the electrostatic charge picked up the fibre from moving belts of dried tea. The process was patented by Brooke Bond and adopted universally.[18] Meanwhile, the old practice of collecting the sorted tea grades into large floor piles was replaced by introducing banks of graded silos which discharged onto belts, with much reduction of dust, as well as a time saving. In the late 1960s, traditional tea chests made from plywood were replaced by cardboard cartons and, after the introduction of container transport, by multiwall paper sacks.

The more that the transformation of harvested green leaf into black tea ready for packaging could be conceptualised as a continuous process of chemical changes, the more a tea factory could be envisaged as a chemical plant, where sophisticated control of temperature and timing could be applied at every stage in order to monitor the underlying transformation. Many permutations of the simple sequences have been tested, including introducing more than one stage of rolling, or cutting or fermenting with differing volumes of leaf – all in the search for better manufacturing quality, in combination with fuel and labour economies. The point has been well made that the advance of scientific understanding of wine making in California and Australia has transformed the wine industry without removing the ultimate judgement of the palate, and that a similar revolution has been under way with tea. One consequence has been marked increases in proprietary knowhow and the guarding of intellectual property. Nevertheless, traditional methods (orthodox tea, as it is called) are still to be found at both ends of the quality spectrum: in manufacturing high-value special teas, especially in north India and Sri Lanka, and in small privately owned tea estates.[19]

A tea factory needs a fuel supply to generate hot air for the drying machines and this was supplied by the estate's own timber plantation. In Kenya the preferred tree was fast growing eucalyptus species and several factories have redundant railway locomotives as their boilers.[20] The 1970s oil crisis arrested any enthusiasm for oil fired boilers (and severely inconvenienced KTDA, which had gone down that route), and the preference now is strongly for wood fuel. The management of the timber plantation is therefore also a focus for applied research: the more productive is the plantation, the less land is taken away from growing tea. The rule of thumb is that 1ha of eucalyptus is required for every 3.5ha of tea, and the trees are coppiced on a 10 year cycle and replanted after 30 years.[21] In the 1990s James Finlay achieved a 60 percent increase in timber yield from better plantation management to yield 600 cuft/ha, with a consequent freeing up of land. It might have been thought that timber supply could be outsourced, but the Kenya timber industry had been plundered and riddled with corruption in the Moi era with the paradoxical result the tea companies were finding a profitable market for any surplus timber.[22]

This brief review of innovation and applied science has revealed the multiple influences at work in the industry. Innovation from practical engineers fostered factory improvements in withering, crushing, tea grading and especially in the CTC revolution. Labour cost pressures have driven the application of mechanical harvesting over hand plucking. Applied research has brought startling benefits in weed control, fertiliser application, vegetative propagation of selected varieties and plant breeding. Customer convenience stimulated the tea bag revolution and the development of new products of instant tea and ready-to-drink tea. The more sophisticated and discriminating management of the tea plant itself and of the manufacturing process has led to a new generation of expertly trained supervisors and managers that is far removed from the traditional caricature of the expatriate tea planter, sitting on his verandah at the end of the day, pink gin in hand. Perhaps the most startling transformation of all has been Kenya's demonstration that high quality tea can be successfully grown on small family farms and integrated into a successful export industry.

NOTES TO CHAPTER 12

[1] This chapter is based substantially on interviews with current and retired executives of James Finlay and Brooke Bond in 2003 and 2004, with further details recorded in the interview notes and memoirs in MP/IM/1/2/3/4 and in the narrative Research Diary PM/IM/5.

[2] MP/IM/5, Interview Joshua Langat, 6 August 2003.

[3] In 2004 I visited tea estates in western Uganda that had been abandoned during the civil war some 20 years previously and the tea had grown into forest trees. They were in the course of being reclaimed and had been cut down again to three feet and leaf harvesting was resuming.

[4] From the outset of smallholder tea growing, the Agricultural Department had the authority to insist on a strict application of the 'two and bud' plucking standard, with ruthless rejection of substandard plucking at the weighing stations and this was successfully embedded in the culture of KTDA. This was clearly in evidence on my visit to Nyeri over 40 years later, in 2003, where, however, there were buyers standing by to purchase the rejected leaf and drive it across the Rift Valley to independent factories. Small family tea holdings had an incentive to pluck consistently, but the widespread employment of casual labour on the larger holdings was a risk.

[5] MP/IM/4/2, Interviews C Tyler, May 2003. MP/IM/3/2, L Stone-Wigg, 17 May 2003

[6] MP/IM/5, Interview S Musongo, 5 August 2003

[7] By way of illustration, the following tabulation of daily wage rates between 1988 and 2003 from the interview with S Musongo shows a sixfold increase over 15 years.

[8] Interview Stone-Wigg, ibid.

[9] MP/IM/4/6, Interview, T Brazier, June 2003.

[10] Interview, C Tyler, ibid.

[11] MP/IM/5, Interview, N Paterson, 13 August 2003. He was a James Finlay man until retirement but, at the time of interview, was the appointed visiting agent to Brooke Bond.

[12] Interview J Langat ibid., who was in charge of research at James Finlay. Soy at Chemamul estate had a field that yielded 8,000 kg/ha one year.

[13] PM/IM/5, Data provided by the manager of Kimugu estate 5 September 2003.

[14] PM/IM/5, Interview G Tuwei, 4 September 2003, who was in charge of plant breeding at Brooke Bond.

[15] The criticism was that the Tea Research Foundation's policy direction was not related closely to industry concerns. At the time of my visit there were no current science journals in the library. Interviews with Langat, Awuor, Tuwei and Onsando.

[16] With Brooke Bond, the replanting programme was running at $1 million per annum, or approximately two percent of the tea acreage. PM/IM/5, Interview, R Fairburn, 4 September 2003. At the time of its investment, there was a disposition by Unilever to have a much faster programme, based on its palm oil experience. Cf. Interview, T Brazier.

[17] Interview C Tyler, ibid.

[18] Ibid.

[19] James Finlay have retained one factory in Kericho at Saosa to manufacture tea by traditional methods, largely for a single speciality buyer in Germany.

[20] MP/AH/4, House magazine, Vol 24/No 3/Autumn 1988.

[21] MP/AH/4, House magazine, Vol 29/No 2/ Autumn 1994.

[22] MP/IM5, Interview, John Othira, 11 August 2003.

13

ENGAGING WITH
THE GLOBAL MARKET

When tea was first shipped in quantity to Europe in the 18th century and its consumption spread beyond the caravan and shipping routes of East Asia, it could be claimed that a global consumer product was being established. Tea became embedded in the cultures of Europe, Russia and throughout the Islamic world. India not only became the leading producer for export, but the product became widely adopted domestically as a staple drink. The 19th century saw its further penetration in the Americas and Africa, and the phenomenon of Kenya's progressive emergence as the leading exporter to world markets. Societies have taken to the drinking in contrasting ways, ranging from the rituals of the tea ceremony to the office and factory tea break and market tea stalls. In their different ways, much of the world has benefitted from tea's unique combination of a mild caffeine stimulant, the safety factor of a beverage using boiled water, and a more general acceptance that its properties are beneficial to health.

During the second half of the 20th century the supply of tea to world markets tended to exceed the growth in consumption, as new tea-growing countries came onto the scene from as far apart as Vietnam and Argentina. But consumer tastes were also changing, especially in the major UK market, where per capita consumption of tea fell from around 8.5 lbs per head in the late 1940s to 5.1 lbs at the end of the century and to 2.4 lbs 15 years later.[1] As a consequence, there has been a statistically weak price outlook which has kept alive thoughts of managing the supply to world markets, albeit without an actual revival of the ITA. Kenya has participated in the FAO consultations and, unsurprisingly, resisted any proposals to restrain development while KTDA was undertaking its successive development plans. However, at the 2003 FAO meeting, the Tea Board agreed to support a proposal for export quotas and a reduction in export volumes.[2] Individual producer countries have been affected year by year by weather events and political developments, so that the actual tea price history has shown considerable fluctuations. In the era of price controls, any divergence between domestic prices and export prices was a source of tension with producers and, as we have seen, contributed to Brooke Bond's loss of its management of the Pool. A brief review of the main trends in the world tea trade will help place the Kenya experience in perspective.

A Global Commodity: World Tea Consumption in 2015[2]

	'000 metric tons
United Kingdom	115
Russia and the Commonwealth of Independent States	262
Rest of Europe	136
North America and West Indies	149
Latin America	28
China	1,924
Rest of East Asia	232
India	983
Rest of subcontinent	303
Middle East	495
North Africa	218
Rest of Africa	132
Oceania	17
Total	4,994

Source: International Tea Committee

Between 1950 and 2015, world tea production rose eightfold from 641,000 to 5,285,000 metric tons. At the start of this 65-year period, India and Sri Lanka accounted for two-thirds of world tea production, but at its close for only a little more than a quarter, due to the increasing prominence of four other tea-producing countries: China, Indonesia, Kenya and Turkey. Of these, Indonesia, and especially China, had a lengthy history of tea growing, and by the end of the period China was accounting for 43 percent of world tea production. Over the same time span, Kenya's contribution had risen from 1 to 8 percent. Throughout this period, these six countries have accounted for over 80 percent of world tea production, although a host of other countries now grow tea and four of them, of which the most dynamic has been Vietnam, contribute between 1 and 3 percent.

World Tea Production 1950–2015

'000 metric tons

	1950	1960	1970	1980	1990	2000	2010	2015
India	278	321	419	570	720	846	966	1,209
Sri Lanka	143	197	212	191	234	307	331	329
China	63	136	136	304	540	683	1,475	2,249
Kenya	7	14	41	90	197	236	399	399
Indonesia	35	31	44	99	145	157	151	133
Turkey	–	6	33	96	127	131	231	259
Vietnam	–	5	6	22	32	70	175	170
The rest	115	223	361	476	528	549	553	557
Total	641	933	1,252	1,848	2,523	2,923	4,281	5,285

Source: International Tea Committee

In the 18th century the world tea trade was entirely accounted for by China and this remained so until its monopoly was broken by the East India Company; in the 19th century India and Sri Lanka became the leading exporters. The 20th century saw the emergence of other significant producers and the International Tea Committee now tracks some 30 of them in its statistical bulletins. There is a sharp contrast between producing countries where there is a large domestic market for tea and those where production is largely for export to world markets. Both China and India have achieved huge increases in production in the post-war period – India's has increased fourfold since 1960, but domestic consumption has also risen sharply. India now retains some 80 percent of production for domestic use, compared with 40 percent in 1960; China, likewise, retains more than 80 percent of its production. Sri Lanka and Kenya are polar opposites: the former retains only about 10 percent of production for the local market and Kenya less than 5 percent. Of other large producers, Indonesia retains around 20 percent of its crop; Turkey's production is almost entirely for domestic consumption, despite its rapid expansion; and Vietnam has achieved very rapid growth in production since 1990 and retains some 20 percent for domestic consumption.

Tea Retained in Main Producing Countries

'000 metric tons

	1950	1960	1970	1980	1990	2000	2010	2015
India	97	128	218	345	511	642	748	983
Sri Lanka	8	11	4	7	19	27	35	28
China	51	95	95	196	345	456	1,173	1,922
Kenya	2	2	5	–	27	9	42	44
Indonesia	7	10	7	31	34	57	64	71
Turkey	–	6	26	91	99	124	227	253
Vietnam	–	3	4	13	19	8	37	37

Source: International Tea Committee

Against this background, Kenya's meteoric 34-fold increase in tea exports from 12,000 metric tons in 1960 to over 400,000 metric tons has projected it to the position of the world's largest tea exporter. It is followed by China with about 330,000 metric tons and Sri Lanka with about 280,000 metric tons. India remains, of course, a major supplier to world markets, with exports of about 230,000 metric tons, followed by Vietnam at 140,000 metric tons and Indonesia at 55,000 metric tons. Notwithstanding the fact that China and India have absorbed most of their increased production, world exports of tea have increased more than threefold since 1960 to some 1,800,000 metric tons, with Kenya as the largest contributor.

Tea Exports of Main Producers

'000 metric tons

	1960	1970	1980	1990	2000	2010	2015
Kenya	12	36	90	170	217	441	443
India	193	200	224	209	204	219	225
Sri Lanka	186	208	184	215	280	296	281
China	41	41	108	195	228	302	325
Indonesia	36	37	68	111	106	87	62
Vietnam	2	2	9	14	56	138	133
Argentina	3	19	33	46	50	85	76

Source: International Tea Committee

Whereas, in the 1950s, UK consumers accounted for approximately half of world tea imports, consumption has fallen relentlessly since then to half the earlier level, at around 110,000 metric tons, and the UK now accounts for only some 6 percent of world tea imports. However, Kenya has become the dominant supplier to the UK market, accounting for more than 40 percent of tea imports and displacing Sri Lanka almost entirely. Kenya is also the dominant supplier of tea to Pakistan (78 percent of imports), Egypt (90 percent) and Sudan (90 percent). By contrast, in the important Russian market, Sri Lanka is the principal supplier and Kenya provides only 12 percent. Although the USA is now a large tea market, at 130,000 metric tons, more than half its imports come from Argentina, so that its significance to other exporting countries lies principally in its role in respect of new product innovation. Here, Kenya is fortunate in its Unilever connection, since its Lipton brands compete strongly in that market.

Kenya's Share of Major Markets

	2000	2010	*% of Imports from Kenya* 2015
UK	41	61	43
Pakistan	52	63	76
Afghanistan	50	57	76
Egypt	68	80	87
Sudan	50	57	76
Commonwealth of Independent States	–	25	36

Source: International Tea Committee

From a post-war position where virtually the whole of the Kenya crop was exported to the UK, this market now only takes around 10 percent and the destination of exports is dominated by North African and Middle East markets, notwithstanding economic and political instability in those regions. The case of Pakistan is particularly interesting. High nominal import duties had led to elaborate smuggling networks and payment arrangements; one consequence was that Afghanistan featured as a major importer of Kenya tea in the trade statistics for many years, until the situation was corrected recently. Pakistan and Egypt between them account for some 50 percent of Kenya tea exports and six country destinations account for three-quarters of exports, or 85 percent if exports to the large entrepôt blending centre of Dubai are included.[3]

This heavy concentration on a few markets could be considered a vulnerable feature. Other countries in the Middle East region are large tea consumers: Iran, Iraq and Turkey especially, but achieving market penetration is not straightforward. Establishing wider recognition and acceptance for Kenya teas will probably require industry-sponsored (and even government-assisted) promotion, as has been the case in Sri Lanka for many years. The marketing challenge for Kenya is to enlarge its tea exports to the 13 countries that take more than 1,000 metric tons a year, but together only account for some 13 percent of its exports at present.

Kenya's Tea Export Destinations

'000 metric tons

	1980	1990	2000	2010	2015
UK	54	71	54	73	45
Pakistan	9	42	58	76	116
Afghanistan	–	–	13	33	42
Egypt	3	19	43	75	77
Sudan	1	6	8	25	20
Commonwealth of Independent States	–	4	–	23	33
UAE	–	1	–	–	38
Subtotal	67	143	176+	305+	371
14 countries one ton plus					58
The rest	23	27	41	136	14
Total	90	170	217	441	443

Source: International Tea Committee

Kenya escaped the post-war fashion for marketing board price stabilisation funds because production was in the hands of large estates at the time; however, as we have seen, it was subject to price control in the domestic market for many years. The ending of bulk purchase by the Ministry of Food and the reopening of the Mincing Lane auctions in London brought Kenya into competition with India and Sri Lanka for the first time. Its reputation as a low-quality volume producer suited Brooke Bond and its strong brands. The revolution of brewing tea in tea bags rather than using loose-leaf spoonfuls favoured Kenya teas. Perhaps affected by nationalisation developments, Sri Lankan producers were slow to adapt to the new consumer convenience and Kenya tea became the product of choice for UK packers, a position which it has substantially retained. Other developments in the subcontinent also had favourable consequences for Kenya tea. The war of independence in 1972 resulted in Bangladesh cutting off the supply of tea to its main market in west Pakistan. The void was filled by Kenya and Sri Lanka until a political fallout in 1975 left Kenya as its main supplier. In 1982, India banned the export of tea for domestic reasons, which lost it the lead position in the UK market, to Kenya's great advantage.

A key step in raising the profile of Kenya teas was the establishment of its own tea auctions in Nairobi in 1956. East Africa Coffee Plantations, which also grew tea in Nandi, requested Dalgety & Co to arrange an auction in Nairobi of its crop. All the members of the Mild Coffee Trade Association were founder members of the East Africa Tea Trade Association, whose first auction was held in November 1956, with offerings from Uganda and Tanzania as well. Brooke Bond refused to participate in the auction and members of the Pool were disbarred from using the auction at first. Holding an auction in Nairobi had the immediate attraction that growers received payment significantly sooner than with tea consigned to London.[4] The auction initiative was a success and this was reinforced by its transfer to the export port of Mombasa in 1969. By this time, Brooke Bond was ready to play an active part in the auction. Kenya hosted an international convention in 1971 (the first of many) to underline its emergence as a trading hub and Mombasa became second only to Colombo as a tea auction centre. This was further enhanced in 1992 by the move to US dollar contracts, which has not been followed by either Colombo or India. Mombasa has now become the auction centre for all African teas, from Malawi, Tanzania, Uganda, Rwanda and other producers. The duty-free allowance for teas imported for blending and re-export has further enhanced the tea-blending operations of James Finlay and others. The historic Mincing Lane auctions closed in 1998 as a result of the success of those in producer countries.[5]

The alternative to selling production at auction is by private treaty sales with major buyers and the overseas marketing arms of some producers. Practice varies quite widely. KTDA sells virtually all its export tea through the Mombasa auction, which is a reflection of the absence of close institutional links overseas. James Finlay sells up to three-quarters of its exports through the auction, Linton Park around a half, Brooke Bond only a quarter and George Williamson hardly any at all as production is consigned to its own packing and marketing operation in England. Unilever has its main blending and packing centre in Dubai and so Brooke Bond's orientation is to that destination. In the case of James Finlay, its trading arm in London receives weekly airmail samples and has a window of opportunity to purchase the factories' production for its private treaty customers before it goes to auction in Mombasa. It pays a premium on top of the auction price (but avoids auction commission) and pays cash against documents, which gives Kericho a faster cash flow than the auction procedures.

There are around 50 buyers at the weekly auctions, with strong representation from buyers from Pakistan and Egypt, where deep connections have been established. In a market tending to oversupply, and with weak demand in its most traditional market, there is a strategic imperative to foster new products and new markets and to capture any premium for quality. Kenya has been fortunate in having industry leadership that was well aware of these possibilities. There are several paths to the goal of premium prices for an otherwise standard commodity. The first is to raise the standards of cultivation and manufacture, which is reviewed more fully in chapter 12. Here it may be noted that perhaps the single most important contribution in Kenya was the insistence by the colonial Department of Agriculture from the outset that smallholder tea should be a quality product. This was inculcated in the new farmers at a time when the estate sector was still aiming at volume before quality. When KTDA tea came onto the market it had the distinction of capturing better prices than that from the estates. Linton Park led the way in Nandi in driving up quality in the estates sector and this has been followed by the others.

A second path to pricing advantage lies in product differentiation and in the application of modern science to refine the consumer offering. A suggestive comparison would be that the estate companies are engaged in a quality transformation analogous to the scientific revolution in the wine industry that was led by Australia and California. This leads to the establishment of tasting reputations for individual estates and to manufacturing methods that can be marketed to discriminating buyers and to meet particular cultural preferences. James Finlay's science group works with beverage firms to refine their products, while the giant Unilever can link its consumer expertise with research at its own estates in Kenya to refine its offerings under the Lipton brand. Linton Park has not chosen to be involved in these downstream refinements. Apart from the quality of traditional black tea to capture pricing premiums, the Kenya industry is at the forefront of new product innovation through the leadership of James Finlay and Unilever.

James Finlay took an early interest in the concept of instant tea in the 1950s. The product had got off to a bad start with consumers with Nestea, which included milk powder. An estate engineer in India began to experiment with a powder concentrate soluble in cold water and was transferred to Kericho to develop a production unit. To the curious, it was said that Finlay's were building a kipper factory and the name stuck. A chemist was recruited who turned out to be something of an imposter, but not before the project was transferred to India for full-scale production in a joint venture with Tata. In 1966 the new factory failed to produce a satisfactory product and James Finlay turned to the professor of agriculture at Cambridge for help. The outcome was the recruitment of a research chemist who had been working in Brazil and the production problem was resolved. The instant tea was aimed at the US market and purchased by the big beverage companies, including Lipton, who had developed an identical product. After James Finlay sold its Indian tea interests to Tata in 1974, attention turned to the mothballed kipper factory in Kericho, which was recommissioned. This coincided with Cadbury losing its factory for making hot-water-soluble instant tea in Uganda during political troubles there, and the kipper factory stepped in and switched to this product. In the meantime, Brooke Bond had developed an instant tea product at a pilot plant in Wales (hot-water-soluble freeze-dried tea) and decided to build a production unit in Kericho using green leaf, in conjunction with Australian associates. However, the business model was based on an input price for green leaf as for that with CTC black tea, which undermined the venture's financial viability so that the partners pulled out and the factory was cannibalised. At this juncture, James Finlay purchased much of the plant, including a 100-foot spray drier. Subsequently, Unilever has built a new instant tea factory.

Both tea and coffee are known for the stimulant effect of the alkaloid caffeine, but the caffeine content of tea is about half that of coffee and its stimulant effects are correspondingly milder.[6] James Finlay was an early pioneer of decaffeinated tea in the 1980s, working with scientists at Strathclyde University in Glasgow. A pilot plant, using a methylene chloride solvent, fell foul of a US ban on this solvent in food products (although not in Europe) and it then had to be adapted to use ethyl acetone as an acceptable alternative for that market. The American market has continued to be the main arena for the development of new tea products. Recent attention has been on ready-to-drink teas in order to compete in the soft drinks market where, to Kenya's advantage, Unilever is prominent with the Lipton brand.

The Japanese taste for its special green tea has been a challenge to which both James Finlay and Unilever have responded, with both instant tea and ready-to-drink products. There are wider implications for the health reputation of tea, arising from research into the antioxidant properties of the flavonoid content of tea. In the West, the focus has been on tea's role in reducing cholesterol levels and its contribution to the control of coronary heart disease. In Japan, there has also been wide acceptance that the green tea that is consumed there makes a positive contribution to reducing body fat, and therefore to weight loss. In an increasingly health-conscious world, a fresh dimension for drinking tea has been opened up, although research insights have yet to be sufficiently strongly endorsed for them to feature in mass promotions to consumers.

A third path for capturing added value for Kenya tea producers lies in engaging in the sophisticated and competitive art of consumer marketing. This lies behind the efforts, already referred to, to achieve recognition for tea made from individual estates and to adapt to the preferences of individual consumer markets. There is a wider question of how best to promote the reputation of Kenya tea as such.

Sri Lanka has been rather successful in establishing a generic reputation for Ceylon tea. Incentives have been provided to encourage locally based packing and branding. There is perhaps a challenge here for KTDA factories and for Ketepa, especially vis-à-vis regional and Middle East markets. More generally, as a beverage, tea has lacked the single-minded global promotion that has driven the success of Coca-Cola and Pepsi, and has not had the benefit of the international retail presence that coffee has had from Starbucks. It remains to be seen whether Unilever's successful development of ready-to-drink tea and its Lipton brand has this potential.

NOTES TO CHAPTER 13

[1] MP/Library/14 and 15, *World Tea Statistics 1910-1990*, and *Annual Bulletin of Statistics 2018*, International Tea Committee. Privately Published in 1996 and 2018.

[2] MP/IM/2/7, Research Diary, Interview with Tea Board chairman, September 2003.

[3] The 'Big Seven' export destinations, in descending order, are Pakistan, Egypt, the UK, Afghanistan, the UAE, Russia and Sudan. The smaller markets with potential are Yemen, Poland, Iran, Somalia, Nigeria, Ireland, USA, Saudi Arabia, Turkey, the Netherlands, Japan and Oman. The concentration in the Middle East is noteworthy.

[4] MP/Sales/3/7 and 8, *East Africa Farmer & Planter*, November 1956 and circular issued by the EA Tea Trade Association, November 1956.

[5] The East African Tea Trade Association produces an informative document on the operations of the Mombasa auctions.

[6] MP/Lib/29, *Tea and Health: A report on the influence of tea drinking on the nation's health*, The Tea Council, 1996. Contribution by Professor R Walker, p.12.

SOURCES AND BIBLIOGRAPHY

In 1956 and 1957 I was able to carry out research in East Africa with funding support from a Rockefeller Foundation grant to Nuffield College and from the Tea Board of Kenya. I had access to three main archive sources:

- File registries in the government departments of Agriculture and Labour in Kenya, Uganda and Tanganyika. National archives were later established in these countries after independence in the 1960s, but not all departmental records were transferred to them.
- Company records kept by the Kenya subsidiaries of James Finlay – especially relating to the early years, and of Brooke Bond, especially regarding the history of the Pool.
- The files of the Kenya Tea Growers Association, especially relating to labour conditions, the International Tea Agreement, and price control.

Manuscript copies were made of relevant documents from these sources.

As a member of the Board of the Commonwealth Development Corporation in the 1990s I was given access to its archive in order to write a history of the corporation to coincide with its 50th anniversary in 2000. This was published as *The Development Business – A History of the Commonwealth Development Corporation.*

Palgrave, 2001. Most of the corporation's historic records had been transferred onto microfilm, from which I was able to make copies of key documents. The reading equipment broke down and was not replaced and so I hired a machine to complete my work. CDC was subsequently broken up and moved offices and the microfilm records have been lost.

In 2003 and 2004 I made two research visits to Kenya and was given generous interview access, especially by James Finlay and Unilever, as well as with a wide range of industry executives and retired managers. This was supplemented by further interviews back in the UK. The outcome is a collection of individual interview notes and a consolidated research diary recording the remaining interviews.

During the 1970s a number of retired Brooke Bond managers recorded brief recollections of their time in Kericho, and I was given a set of these memoirs. Separately, two brothers had recorded their recollections of the 1920s and I was given a copy when interviewing the widow of one of them. She had subsequently married another early planter who had set down his account of the first settlers in Kericho, and I was given a copy.

Hugh Thomas, one of the first planters in Kericho, wrote a history of the African Highlands Produce Company up to the 1950s for private publication by James Finlay. He gathered recollections from some of the early tea planters and compiled a note on the origins of the industry. He also recorded a memoir and I have copies of these documents. Thomas rose to be Superintendent and he eventually became my stepfather.

The common feature of the above sources is that many of the original documents may no longer exist, or are not readily accessible to a modern enquirer. However, arrangements have been made to deposit the archive material that has been the source of this book with the Bodleian Library in Oxford with the classification references set out below.

McWilliam Papers: MP
Interviews & Memoir Transcripts: MP/ IM

Brooke Bond Memoirs: IM/1
MP/IM/1/1	Chris Tyler, May 2003
MP/IM/1/2	Oliver Brooke, 27 January, 4 February 1999
MP/IM/1/3	Tom Grumbley, 20 February 1998
MP/IM/1/4	Michael Gerken, 22 June 1999
MP/IM/1/5	Peter Kennedy, 8 January 1998
MP/IM/1/6	Peter Knight,16 & 17 February 1999
MP/IM/1/7	Tony Monkhouse
MP/IM/1/8	Musa Sang, 26 November 1998
MP/IM/1/9	Hugh Morton, 11 December 1997
MP/IM/1/10	Bridgette Ball, 24 January 1998
MP/IM/1/11	Joyce Pickford, 21 January 1978 and 1971 (2 docs)

Industry Interviews: IM/2
MP/IM/2/1	Robin Harrison, Thompson Lloyd Ewart, 22 May 2014
MP/IM/2/2	Norman Wilson, African Tea Brokers, 21 August 2003
MP/IM/2/3	David Venters, Combrok, 21 August 2003
MP/IM/2/4	Richard Kimoli, CDC, 17 February 2004
MP/IM/2/5	Gatabaki, CDC, 17 February 2004
MP/IM/2/6	Francis Wacima & Jean Cliffe, 28 August 2003
MP/IM/2/7	Stephen Nkanata, Tea Board, 1 September 2003
MP/IM/2/8	Tiampati, Katepa, 5 September 2003
MP/IM/2/9	Nick Kirui, KTGA, 17 February 2004
MP/IM/2/10	Denton Vorster, Eastern Produce, 8 February 2004
MP/IM/2/11	Malcolm Perkins, Linton Park, 27 April 2004

James Finlay Interviews: IM/3
MP/IM/3/1	Nev Davies, August 2003, February 2004
MP/IM/3/2	Lindsay Stone-Wigg, 17-19 May 2003
MP/IM/3/3	Rupert Hogg, July 2003

MP/IM/3/4   Bill Eyton, July 2003
MP/IM/3/5   Richard Darlington, J F Overseas, July 2003
MP/IM/3/6   Peter Robertson, 14 February 2004

Brooke Bond Interviews: IM/4
MP/IM/4/1   Kericho interview, 4 & 5 September 2003
MP/IM/4/2   Chris Tyler, 6 May 2003
MP/IM/4/3   Oliver Brooke, 9 October 2004
MP/IM/4/4   Joyce Pickford, 8 & 18 February 2004
MP/IM/4/5   John Popham, June 2003
MP/IM/4/6   Tom Brazier, June 2003

Research Diary & Interviews: IM/5

International Tea Agreement: MP/ITA
Department of Agriculture files in Kenya, Uganda and
Tanganyika and KTGA
MP/ITA/1/1   February 1933 – August 1939
MP/ITA/1/2   January 1937 – May 1948
MP/ITA/1/3   Uganda, Belgian Congo, Mozambique
MP/ITA/1/4   ITA Chronology and McW Review Note
MP/ITA/1/5   FAO Commodity Review 1961-70
MP/ITA/1/6   FAO Intergovernmental Group on Tea 1973–99

Labour: MP/Labour
MP/Labour/1/1   KTGA, Labour Dept, Official Reports 1921–55
MP/Labour/1/2   Uganda
MP/Labour/1/3   McWilliam Reflections 1958
MP/Labour/1/4   Wage Rates 1976–2003
MP/Labour/1/5   Brian Cranwell: Industrial Relations

Early Days in Kenya: MP/EDK
MP/EDK/1/1   John Wilson
MP/EDK/1/2   H Thomas: Notes on the History of Tea
                Growing in Kenya. 1950?
Letters: Orchardson, Caine, Matthews

Thomas Notes
East African Standard, 1 October 1929
Imperial Institute Report, 26 March 1909
Limuru Farmers Pamphlet, 1916
English Scottish Trust, 1948
MP/EDK/1/3   Buret Tea Company statistics, 1927–43
MP/EDK/1/4   Agricultural Department Reports, 1924–55

Kenya Tea Growers Association: MP/KTGA
MP/KTGA/1/1   Censuses of Labour and Dependents 1980–2002
MP/KTGA/1/2   Membership and holdings 1976–2002
MP/KTGA/1/3   Tea Board Licence Holders 1959
MP/KTGA/2/1   KTGA Rules 1954
MP/KTGA/2/2   District Planting 1938–43
MP/KTGA/2/3   Chronology 1931–34
MP/KTGA/2/4   Minutes 1932–51
MP/KTGA/2/5   Topics:
1931–35 Excise Duty
1933–37 Tea Cess
1933–40 Lumbwa Road
1939 Railway Rates
1945 Eucalyptus Snout Beetle
MP/KTGA/2/6   E.A. Tea Association 1935–44
MP/KTGA/2/7   Kericho Population 1944
MP/KTGA/2/8   Origin of Tea Research Institute
MP/KTGA/2/9   Estate and District Production, various

Tea Marketing – The Pool: MP/Sales
MP/Sales/1/1   Brooke Bond files 1935–38
MP/Sales/1/2   B. Bond files 1938–42, War contracts
MP/Sales1/3   B. Bond files 1941–44, Pool Reports and
                 D.S.McW Review 1944
MP/Sales/1/4   Uganda Tea Association 1937–44
MP/Sales/1/5   Tea Cess Board
MP/Sales/1/6   Consumption within E. Africa 1932–39
MP/Sales/2/1   B. Bond files 1945–55, Pool Reports

MP/Sales/2/2	B. Bond files 1945–52, Price Control
MP/Sales/2/3	B. Bond files 1947–56, Advisory Committee
MP/Sales/2/4	M McW. Review of Pool, 1957
MP/Sales/3/1	Tea Controller statistics
MP/Sales/3/2	Tea Chest 1956: Sales Organisation
MP/Sales/3/3	Katepa
MP/Sales/3/4	Estate Production Costs 1939–45
MP/Sales/3/5	Consumption per head/district 1949–56
MP/Sales/3/6	Pool Valuations 1948–57
MP/Sales/3/7	Tea Auction in Kenya 1956
MP/Sales/3/8	East African Tea Trade Association

Limuru: MP/Limuru

MP/Limuru/1/1	Early tea farmers: output 1936–45
MP/Limuru/1/2	Wacima interview and sales to elite
MP/Limuru/1/3	Sorrenson note and land grievances

African Tea: MP/African Tea

MP/African Tea/1/1	Agriculture Dept files K,U,T, 1945–57
MP/African Tea/1/2	Central Province Tea Board 1957–58
MP/African Tea/1/3	McWilliam memos 1957
MP/African Tea/1/4	G Gamble. *Expansion of Tea Planting in the Central Province, Mimeo.* 7 August 1958

Smallholder Tea & KTDA: MP/CDC
Copies taken from Commonwealth Development Corporation archive 1997–99

MP/CDC/1	Smallholder Development 1958–1960
MP/CDC/2	Special Crops Development Authority 1960–64
MP/CDC/3	Kenya Tea Development Authority 1964–67
MP/CDC/4	Kenya Tea Development Authority 1968–69
MP/CDC/5	Kenya Tea Development Authority 1970–73
MP/CDC/6	Kenya Tea Development Authority 1974–80
MP/CDC/7/1	Kenya Tea Development Authority 1980–89
MP/CDC/7/2	Nyambeni Tea Co.
MP/CDC/7/3	Chris Walton Reflections, 2006

African Highlands Produce Co./ James Finlay/Swire: MP/AH
MP/AH/1	Office correspondence 1925–53
MP/AH/1/2	Labour force statistics 1925–43
MP/AH/1/3	Peter Robertson interview 1957
MP/AH/2	Company accounts: 5-yearly 1970–95
MP/AH/3/1	Crop statistics 1964–2002
MP/AH/3/2	Kaproret and Kapsongoi Estate costs 1970–2002
MP/AH/3/3	All estate costs 2001
MP/AH/3/4	Estate sales 1982
MP/AH/3/5	Medical Expenses 1990–2002
MP/AH/4	Finlay's House Magazine – Select Issues 1988–2003

Kenya Tea Company/Brooke Bond/ Unilever: MP/BB
MP/BB/1/1	Production Costs 1936–43/5
MP/BB/1/2	Return on Capital 1925–57
MP/BB/1/3	Monthly Estate Production 1996–2002
MP/BB/1/4	Weekly Return 2003
MP/BB/1/5	Tea Chest articles 1954–56
MP/BB/1/6	Kerenga-Chebown/Kimulot Transaction
MP/BB/1/7	Timbilil Purchase
MP/BB/2	Estates Circulars 1922–40
MP/BB/3	Brooke Bond Liebig annual accounts 1972–2002

Statistics: MP/Stats
MP/Stats/1/1	Kenya Tea Acreage & Production 1925–56
MP/Stats/1/2	ditto 1963–2002
MP/Stats/1/3	Wilson Smithet Survey 1959
MP/Stats/1/4	Tea Board: Production By Sub-Sector 2013
MP/Stats/1/5	Tea Exports 1925–56; 1970s–2002
MP/Stats/1/6	Tea Export Prices 1956–2009

Tanganyika Tea: MP/Tang
MP/Tang/1/1	Agriculture & Labour Dept. files 1908–50s
MP/Tang/1/2	Notes on Estate Visits 1957
MP/Tang/1/3	Statistics
MP/Tang/1/4	Visit to Theobald/Tatepa 2003

MP/Tang/1/5	Estate Acreages 1986
MP/Tang/2/1	W.G.Dickinson: The Origin of the Tanganyika Tea Company and its Activities 1940–1948
MP/Tang/2/2	Frank Walker memoir

Uganda Tea: MP/Ug

MP/Ug/1/1	Agriculture Dept. files 1933–50s
MP/Ug/1/2	Uganda Company files
MP/Ug/1/3	Ankole Tea
MP/Ug/1/4	Dr Eden Reports
MP/Ug/1/5	Early History
MP/Ug/1/6	Stats
MP/Ug/2	Ruwenzori Highlands Visit 2004

Library: MP/Lib

The publications listed below were printed for a restricted audience and are not widely available.

MP/Lib/1	David Wainwright: *Brooke Bond: A Hundred Years*. Newman Neame. Published privately, c1970
MP/Lib/2	*James Finlay & Company Limited: Manufacturers and East India Merchants*. Published privately, 1951
MP/Lib/3	Hugh O Thomas: *A Brief History of the African Highlands Produce Company Limited.* Published privately, c.1958
MP/Lib/4	Michael Manton: Camellia: The Lawrie Inheritance. Published privately, 2000
MP/Lib/5	Bob Crampsey: *The King's Grocer: The Life of Sir Thomas Lipton.* Published privately, Glasgow, 1995
MP/Lib/6	Joan Karmali: *The Story of Sotik Tea.* Published privately. Nairobi, 1991
MP/Lib/7	Tea Board of Kenya: *Tea in Kenya*. Department of Information, Nairobi 1959. New edition, c.1965

MP/Lib/8	Wilson Smithet & Co.: *Notes on Tea Estates in Africa.* Published privately, London 1954. New edition, 1962
MP/Lib/9	EW Russell, ed.: *The Natural Resources of East Africa. East Africa.* Literature Bureau, 1962
MP/Lib/10	*African Land Development in Kenya 1946–62.* Nairobi. Ministry of Agriculture, 1962
MP/Lib/11	T.H.Hutchinson: *1996 Kenya Up-Country Directory, containing names of some of the expatriates of European origin who have lived in up-country Kenya.* Privately published with separate Index, 1991
MP/Lib/12	*Tea Handbook.* IBRD Commodities and Export Projections Division, 1982
MP/Lib/13	*Tea Estate Practice.* Tea Research Institute of E. Africa, 1966
MP/Lib/14	*World Tea Statistics 1910–90.* International Tea Committee. Privately published, 1996
MP/Lib/15	*Annual Bulletin of Statistics.* International Tea Committee. Privately published, 2018
MP/Lib/16	*Memorandum on the operations of the International Tea Committee.* Privately published, 1946
MP/Lib/17	*Report for the period 1ˢᵗ April 1941 to 31ˢᵗ March 1949.* International Tea Committee. Privately published, 1949
MP/Lib/18	LH Brown: *A Report on the Tea Growing Potential of Kenya* Ministry of Agriculture. Mimeo, 1965
MP/Lib/19	*The Operations and Development Plans of the Kenya Tea Development Authority.* Nairobi, 1964
MP/Lib/20	RJM Swynnerton: *Smallhollder Development: A Summary Paper.* CDC, 1967
MP/Lib/21	RJG LeBreton: *A Study of the Operations of the Kenya Tea Development Authority.* CDC, 1971
MP/Lib/22	D. Sullivan: *A Review of Kenya Tea Development Authority.* IBRD for CDC, 1974

MP/Lib/23	G Lamb and L Muller: *Control, Accountability and Incentives in a Successful Development Institution. The Case of KTDA.* World Bank working paper 550, Sept. 1982
MP/Lib/24	Anne Thurston: *The Genesis and Implementation of the Swynnerton Plan.* Oxford Development Records Project, 1984
MP/Lib/25	MD McWilliam: *The East African Tea Industry 1920–50.* B.Litt. thesis, Oxford, 1957
MP/Lib/26	MD McWilliam: *The Kenya Tea Industry.* East African Economics Review. July, 1959
MP/Lib/27	Ian Q Orchardson. *The Kipsigis.* East African Literature Bureau. Nairobi, 1961
MP/Lib/28	Tea Ordinance 1950 and Tea Rules 1951. Government Printer Nairobi, 1956
MP/Lib/29	*Tea and Health.* The Tea Council, 1996
MP/Lib/30	*Statistical Abstract* 1986, 1991, 1994, 2002, 2012. Kenya National Bureau of Statistics.

Publications Cited in the Text
Official Reports

Crown Grants: Kenya. Cmd 2747, 1926

Report of the Kenya Land Commission. Cmd 4556, 1934

Report of the Employment of Juveniles Committee. Government Printer, Nairobi, 1938

Report of the Commission Appointed to Enquire into the Question of the Introduction of Conscription of African Labour for Essential Purposes. Government Printer, Nairobi, 1942

Report on Visits to India, Malaya and Ceylon with some notes for the guidance of tea planters in Kenya. Government Printer, Nairobi, 1951

Report of the Committee on African Wages. Government Printer, Nairobi, 1954

A Plan to intensify the development of African Agriculture in Kenya. Government Printer, Nairobi, 1954

East Africa Royal Commission 1953–55: Report. Cmd 9475

Liberalisation and Restructuring of the Tea Industry. Sessional Paper
No 2, 1999

Other Publications

Farson, Negley: *Last Chance in Africa.* Gollancz, 1949
Forrest, Denys: *Tea for the British.* Chatto & Windus, 1973
Huxley, Elspeth: *The Sorcerer's Apprentice.* Chatto & Windus, 1948
Langat, SC: *Some Aspects of Kipsigis History Before 1914:*
   E.A. Publishing House, 1969
MacFarlane, Alan and Alice: *Green Gold.* Ebury, 2003
McWilliam, Michael: *The Development Business: A History of the
   Commonwealth Development Corporation.* Palgrave, 2001
Moxham, Roy: *Tea.* Constable, 2003
Sorrenson, MPK: *Land Reform in the Kikuyu Country.* OUP, 1967
Throup, David: *Economic and Social Origins of Mau Mau.* James
   Curreyj, 1987
Wickizer, VD: *Tea Under International Regulation.* Stanford, 1944
IBRD: *The Economic Development of Kenya.* John Hopkins, 1963

INDEX

Lightning Source UK Ltd.
Milton Keynes UK
UKHW020233040820
367650UK00011B/455